HOW TO BUILD

LS GEN IV

Performance on the Dyno

Optimal Parts Combos for Max Horsepower

Richard Holdener

CarTech®

CarTech®

CarTech®, Inc.
838 Lake Street South
Forest Lake, MN 55025
Phone: 651-277-1200 or 800-551-4754
Fax: 651-277-1203
www.cartechbooks.com

Edit by Paul Johnson
Layout by Monica Seiberlich

ISBN 978-1-61325-340-3
Item No. SA395

Library of Congress Cataloging-in-Publication Data

Names: Holdener, Richard, author.
Title: How to build LS Gen IV performance on the dyno / author, Richard Holdener.
Description: Forest Lake, MN : CarTech, [2017]
Identifiers: LCCN 2016043399 | ISBN 9781613253403
Subjects: LCSH: Chevrolet automobile–Motors–Modification–Handbooks, manuals, etc. | Chevrolet automobile–Performance–Handbooks, manuals, etc. | Chevrolet automobile–Parts–Handbooks, manuals, etc. | General Motors automobiles–Motors–Modification–Handbooks, manuals, etc. | General Motors automobiles–Performance–Handbooks, manuals, etc. | General Motors automobiles–Parts–Handbooks, manuals, etc.
Classification: LCC TL215.C5 H65 2017 | DDC 629.25/040288–dc23
LC record available at https://lccn.loc.gov/2016043399

Written, edited, and designed in the U.S.A.
Printed in China
10 9 8 7 6 5 4 3 2

Back Cover Photos

Top:
Cam swaps are one of the most popular upgrades on the an LS3, and for good reason. Tested on this LS3 crate motor with Chevy Performance CNC L92 heads, the Comp cam increased power by as much as 70 hp.

Middle:
An LS3 crate engine served as the test mule, and it was fitted with a set of long-tube headers, a FAST throttle body and XFI management system. The crate motor produced 493 hp and 484 lb-ft of torque with the stock heads.

Bottom Left:
To make a stroker combination even better, you just add boost! This 4.0L Whipple pushed power into the 4-digit range.

Bottom Right:
One potential weakness of the LS3 (as with all LS engines) was the factory valvesprings. Adequate for stock or low-lift applications, the stock springs should be upgraded when swapping cams.

DISTRIBUTION BY:

Europe
PGUK
63 Hatton Garden
London EC1N 8LE, England
Phone: 020 7061 1980 • Fax: 020 7242 3725
www.pguk.co.uk

Australia
Renniks Publications Ltd.
3/37-39 Green Street
Banksmeadow, NSW 2109, Australia
Phone: 2 9695 7055 • Fax: 2 9695 7355
www.renniks.com

CONTENTS

ABOUT THE AUTHOR

It is hard to believe that all of this dyno madness started with a phone call to then-editor of *Turbo* magazine, Kipp Kington. Who would have guessed that a simple story about his adventures at the Silver State Open Road race would send Richard Holdener on a completely different career path? Much to the dismay of his parents, the advertising degree was not utilized in some fancy office on Madison avenue. Instead, he decided to follow his passion for all things automotive.

First as a reader then as a writer, Holdener was fascinated by genuine testing. Forget all the advertising (and now Internet) hype about a product; he said "Let's put it on the dyno and see how she does!" For his first day on staff, he was the guy pushing for dyno testing and acceleration testing the performance potential of products. As you might imagine, this type of verification was met with some resistance from advertisers. According to Holdener, "The readers deserved to know what works and what doesn't."

If you want to find Richard Holdener, look no further than on the dyno at Westech Performance. For more than 25 years, he has worked as a technical editor for a wide range of automotive magazines, both on staff and in a freelance capacity. Richard has specialized in direct back-to-back dyno testing of performance products. He is the crazy guy who compared every factory LS cam, 20 different LS intake manifolds, and dyno (and flow) tested no less than 30 different LS cylinder heads. He is also the guy who performed the Big Bang Theory test, where they cranked up the boost on a stock LS short-block to find out how much power it could withstand. He is the reason so many 1,000-hp stock, short-block LS engines exist.

Unfortunately, you can also thank him for the price increase on LS engines from the wrecking yard, as after the results of the Big Bang Theory, guys were swarming the wrecking yards gathering all the LS motors they could to install turbos. Always looking to illustrate what really works, Holdener even went so far as to design his own adjustable intake manifolds for testing (see Chapter X [intake]). Whether it's blown, built, or boosted, everything from stock LS3 crate engines to 8,000-rpm, short-stroke screamers, Holdener has built and tested it on the dyno.

In addition to this (his 10th) book on LS3 and LS7 Performance, Holdener has also written several other books on dyno verification. These include *Dyno-Proven GM LS1 thru LS7 Performance Parts*, *Building 4.6L/5.4L Ford Horsepower on the Dyno*, *High-Performance Ford Focus Builder's Handbook*, *How To Build Honda Horsepower,* and *5.0L Ford Dyno Tests*. Holdener currently contributes to all the major automotive magazines, including *Hot Rod*, *Car Craft*, *Super Chevy*, *Muscle Mustangs & Fast Fords*, *Power & Performance News*, and *GM High Tech*.

INTRODUCTION

Even the competition has to agree that Chevy's LS engine family is more than just a worthy successor to the original small-block; it's one hell of an engine. The Blue Oval boys were jumping up and down about their new 5.0 Coyote, but (as usual) they were still behind the eight ball in terms of displacement and power output. Although the new four-valve 5.0 modular engine offered reasonable high-RPM power, it was decidedly lacking in low-speed power compared to the LS3. Credit the extra displacement offered by 6.2 liters of displacement (7.0 liters on the LS7) for all that wonderful torque.

High-RPM power is all well and good, but the vast majority of spirited (street) driving comes lower in the rev range. Besides, in the LS (3 or 7) there is no choice between low-speed and high-RPM power, as the GM engines offer both. Toss in the fact that the LS3 and LS7 featured lightweight, all-aluminum construction, composite intakes, and even variable cam timing, and you have a traditional small-block with all the technology of a DOHC Ford engine, without the penalties in size, weight, and complexity.

In the original muscle car era, it took a big-block to muster power ratings that exceeded 400 hp and a like amount of torque, and those old power ratings were gross and not net! The LS3 and LS7 made this a good time to be a Chevy owner, but this book is all about how to make a good thing even better.

The LS engine family has evolved constantly to keep General Motors ahead of the competition. The original LS1 was a solid step above the LT-1, just as the LT-1 easily eclipsed the performance of the previous L98 TPI engine. The LS3 followed the 5.7 LS1/LS6 and 6.0 LS2 performance engine configurations.

Starting with an increase in displacement, the LS3 checked in at 6.2 liters versus the previous 6.0-liter LS2 combination. This came courtesy of an increase in bore from 4.00 inches (in the LS2) to 4.065 inches (the two shared the same stroke of 3.622 inches). The increase in bore size increased displacement and airflow because head flow increases with bore size. The LS7 took this one step further by combining a 4.125-inch bore with a 4.0-inch stroke.

The revised cylinder head(s) that replaced the cathedral-port design with a more conventional rectangular port helped make the LS3 and LS7 serious small-blocks. Tested on the flow bench, production LS3 heads flow as much as 315 cfm right out of the box (350 cfm for the LS7 heads). Those are flow numbers reserved for race heads not long ago, and it takes pretty serious 23-degree small-block (or even cathedral-port LS) heads to reach the flow numbers offered by the stock LS3. Despite flow figures that suggest supporting more than 630 hp (I made as much as 690 hp with a set of stock LS3 heads on a 468 stroker), additional flow is available with proper porting or the substitution of aftermarket LS3- or LS7-based cylinder heads.

Stock LS3 and LS7 heads offer massive airflow, and it's one of the major reasons that they respond so well to cam swaps. A cam is really the only thing missing in the LS package (along with valvesprings). One important point to mention regarding testing in this book is that because the stock LS3 and LS7 cylinders heads offer so much flow, you shouldn't expect huge power gains from a head swap, no matter what the flow bench says. If your modified LS3 (or LS7) makes 600 hp with a set of (350 cfm) heads capable of supporting 700 hp, don't expect much of a change when you add heads with (400 cfm) flow numbers that support 800 hp. The problem isn't (likely) the ported heads, but rather the engine. See Chapter 4 to find out how much power ported heads are worth on combinations ranging from a stock LS3 to a 495-inch stroker LS7.

In addition to camshafts and cylinder heads, this book contains separate chapters on nearly every aspect of LS3 and LS7 performance, including intake manifolds (Chapter 1), nitrous oxide (Chapter 7), and even forced induction (Chapters 5 and 6).

Chapter 5 covers all the forms of supercharging, including Roots, twin-screw, and centrifugal superchargers. Chapter 6 on turbocharging covers both single and twin turbo testing. As well as LS3 and LS7 engines respond to camshafts, they respond even better to boost. Using boost from a supercharger or turbocharger, it is possible to increase the power output of your LS3 or LS7 by 50 to 100 percent or more. As illustrated by the test data in the two chapters, boost is simply a multiplier of the original output. Adding a turbo or supercharger to a stock engine results in less of a power gain at any given boost level than adding the same boost to a modified engine. I also cover the results of turbocharged and supercharged cam testing because the specs differ on cams designed for forced induction.

One thing you will find out about the LS3 and LS7 in this book is the relative strength of their intake manifolds. Testing has shown that the factory LS3 intake is very tough to improve upon. It is possible to increase power higher in the rev range (usually beyond 6,500 rpm) with a short-runner intake, but this usually comes with a trade-off in power lower in the rev range. The two tests on the adjustable intake manifolds (mine and the unit from FAST) clearly illustrate this effect on the power curve.

The comparison between single- and dual-plane carbureted intakes shows this as well, as intake manifolds are designed to operate effectively at specific engine speeds. Short-runner (or single-plane carbureted) intakes should be combined with more aggressive cam timing designed to enhance power production higher in the rev range. By contrast, the factory LS7 intake is very limiting, with significant gains available from an upgrade. Working with intake manifolds are throttle bodies, which offer increased flow. The gains offered by throttle body upgrades increase with the power output of the engine. Tested on a stock engine, a throttle body upgrade might be worth nothing, but tested on an 800-hp combination, it can be worth as much as 50 to 60 hp (especially on a positive displacement supercharged application).

Chapter 7 discusses how nitrous oxide can be applied to any LS combination, ranging from a stock crate engine to a dedicated stroker (including turbo and supercharged combos). The amount of power supplied by nitrous oxide is a function of the jetting, as larger jets allow more nitrous flow. Of course, this must be accompanied by the proper amount of fuel, but nitrous systems offer far and away the most bang for the buck. It is possible to add as much as 250 hp (or more) to your LS for about the cost of a cam swap. Although you make more power with nitrous *and* a cam, every LS owner should experience nitrous oxide once in their life. I have divided the chapter into individual components (i.e., heads, cams, and intakes), but the reality is that the best way to produce optimum power from your LS3 or LS7 combination is with the proper combination of components. The heads must work with the cam timing and intake design to optimize power production in the same RPM range.

Chapter 8 illustrates the testing of combinations designed to work together, ranging from the stock LS3 crate engine to a massive RHS stroker displacing nearly 500 ci.

If you want to know how to make your LS3 or LS7 more powerful with dyno-verified results, you'll find it in these pages.

Intake Manifolds

Whether you have a stock, street, or strip LS application, the intake manifold is one of the three major players in terms of power production. The aftermarket has produced intake combinations for performance LS3 and LS7 applications. Intake designs do more than just allow airflow into the ports; they actually provide a tuning effect that aids in power production over a given RPM range. Not surprisingly, factory LS3 or LS7 intake manifolds were designed with a combination of peak and average power combined with ease of production and even fuel mileage.

The right intake can help you produce impressive power, especially when used in conjunction with the right cam and ported cylinder heads. More than any other single component, the intake manifold (most specifically the runner length) determines where the engine makes effective power. Match the runner length to produce power in the same operating range as the cam profile and you are a long way toward making an impressive LS combination.

For any engine (including LS3 and LS7), intake manifold design may be broken down into three major elements: runner length, cross section (and taper ratio), and plenum volume. These elements are listed in the order they most affect the performance of a given manifold. By this I mean that changing the runner length has somewhat more of an effect than altering the cross section or plenum volume. This is not to say that all of the elements are not important, it is just that proper care should be given to the elements in accordance with their eventual effect on performance. Take note, intake designers often spend countless hours altering the plenum volume in an attempt to change the effective operating range when they should have simply increased (or decreased) the runner length. Also, manifold design is sometimes limited by production capability or rather ease of construction. Building a set of runners with a dedicated taper ratio and a compound curve is difficult, if not impossible, for the average fabricator. Despite the fact that this design produces the best power, it simply isn't going to get produced unless a major intake manufacturer (like FAST, Holley, or Edelbrock) steps up to the cost of such a complex combination.

Fabricated, short-runner intakes such as this unit from Speedmaster are popular among LS enthusiasts, but know that the design lends itself to power production higher in the rev range than the stock (long-runner) LS3 or LS7 design.

The first element in intake design is the runner length. The overall intake runner length actually includes the head ports, but the discussion will be limited to those in the manifold. Fuel-injected intake manifolds seem to be broken down into two distinct groups, long and short. Obviously not very scientific, the terms "long" and "short" do not properly describe intake manifolds. The reason for the long and short designations is that, generally speaking, the longer the runner length, the lower the effective operating rpm. Obviously the opposite is also true because shorter runner lengths improve top-end power. Production LS intake manifolds are typically of the long-runner design to help promote torque production. It is possible to design an intake that offers more low-speed *or* top-end power than the stock LS3 intake, but doing both has proven to be difficult. It should be pointed out that the "ideal" intake design varies with engine configuration as well because the power gains offered by a given design on a stock engine are most likely different on a wilder combination. This is why FAST designed its adjustable LS3 intake manifold to allow adjustment for individual combinations. Since the reflected wave is determined by the cam timing, its initiation point changes with different cam profiles. Thus, changing the cam timing may well require a different intake design.

The next element in intake design is cross section, or port volume. A related issue is taper ratio, but I will cover that shortly. The port volume or cross section of the runner refers to the physical size of the flow orifice. Suppose you have an intake manifold that features 17-inch (long) runners that measure 2.00 inches in (inside) diameter. It is possible to improve the flow rate of the runners by increasing the cross-sectional area. Suppose you replace the 2.00-inch runners with equally long 2.25-inch runners. The larger 2.25-inch runners flow a great deal more than the smaller 2.00-inch runners, thus improving the power potential of the engine. From a reflected wave standpoint, the increase in cross section has no effect on the supercharging effect, but it alters the Inertial Ram and Helmholtz resonance.

Related to the cross section, taper ratio refers to the change in cross section over the length of the runner. Typically, intake manifolds feature decreasing cross sections, where the runner size decreases from the plenum to the cylinder head. The decrease in cross section helps to accelerate the airflow, thus improving cylinder filing, but the real difference is the effective change in cross section brought about by the taper.

The final element of an LS intake manifold is plenum volume. This refers to the size of the enclosure connecting the throttle body to the runners. Typically the plenum volume is a function of the displacement of the engine. Most production intake manifold applications feature plenum volumes that measure smaller than the displacement of the engine (somewhere near 70 percent), but this depends on the intended application. A number of manufacturers have recognized the importance of the plenum volume and incorporated devices to alter the plenum volume to enhance the power curve, but the LS3 and LS7 manifolds rely on a fixed volume.

Contrary to popular opinion, increasing the plenum volume does not increase the air reservoir allotted to the engine as much as it affects the resonance wave. When excited, the area in the plenum resonates at a certain frequency. Changing the plenum volume changes the resonance frequency. The Helmholtz resonance wave aids airflow through the runner (acoustical supercharging). Where this assistance takes place in the RPM band is determined by a number of things but primarily by the plenum volume. The air intake length, inside diameter, and a portion of the cylinder (when the valve is open) are also used to calculate the Helmholtz resonance frequency (and why air intake length and diameter have a tuning effect on the power curve).

LS applications also run very well with carbureted intake systems such as this dual-quad Holley Hi-Ram.

For the ultimate in LS3/ LS7 induction systems, look no further than an individual-runner intake system.

Test 1: Holley Single- vs Dual-Plane Intake on an LS3

When it comes to carbureted engines (including LS), the choice basically comes down to single- or dual-plane. That particular induction argument predates the LS engine family by multiple generations, but carbureted LS owners must ultimately choose. We all know that the LS was originally equipped with factory fuel injection, but MSD made the carb conversion ultra simple. Carb swappers were soon faced with the same induction question that plagued previous small-block Chevy owners. Choosing the proper intake design is critical for maximum performance, but just what defines the term maximum?

In most cases, it doesn't mean peak power, but rather maximized power through the entire rev range. Now throw in things like drivability, fuel mileage, and even torque converter compatibility, and you start to understand the dilemma. You see, despite similar peak power numbers, the two Holley (carbureted) LS intakes tested here offered decidedly different power curves (and likely street manners). We all like to brag about peak power numbers, but the reality is that the vast majority of carbureted LS engines spend *most* of their time well below the power peak. In fact, street engines spend most of their time well under the torque peak and even during hard acceleration, the engine operates primarily between peak torque and peak power.

The choice ultimately comes down to where you value power production. For those new to LS performance (though this applies to every type of V-8 regardless of generation or manufacturer), the intake debate between single- and dual-plane manifolds is a simple matter of operating (engine) speed. The dual-plane was designed to enhance power production lower in the rev range than the single-plane. This simple fact makes the dual-plane ideal for the vast majority of street applications.

Run on the LS3 test engine (with mild Comp cam), the Holley dual-plane produced peak numbers of 544 hp at 6,900 rpm and 471 ft-lbs of torque at 4,300 rpm. After installation of the single-plane intake, the peak numbers changed very little to 552 hp at 7,000 rpm and 463 ft-lbs at 5,200 rpm. Despite minor changes in peak power, the power curves were decidedly different. Check out the curves and decide where you want your LS3 power production.

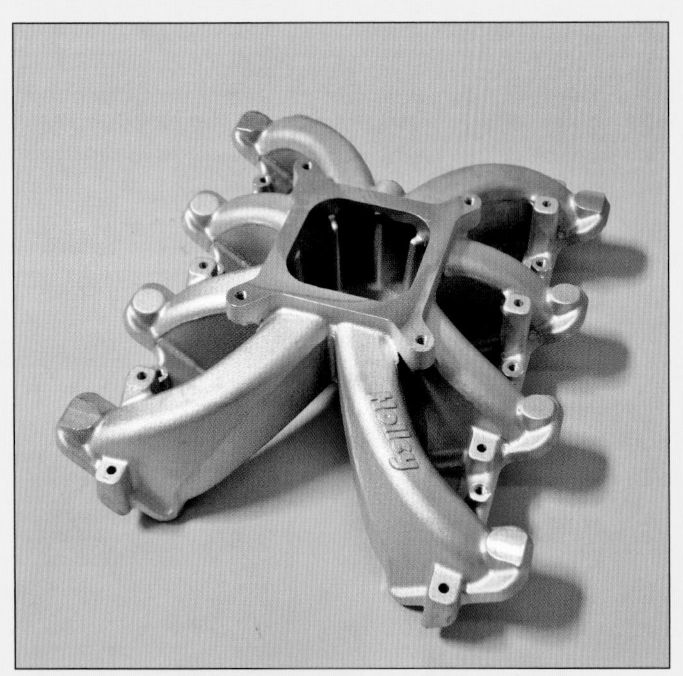

Holley's single-plane intake was designed to optimize power production higher in the rev range than the dual-plane. Just make sure to apply it to the proper combination that can take full advantage of the top-end power production.

Most street and street/strip (carbureted) LS3 applications prefer the dual-plane design because of improved throttle response at lower RPM. The dual-plane was designed to maximize low- and mid-range torque production where it can be enjoyed most often.

Holley Single- vs Dual-Plane Intake on an LS3 (Horsepower)

Holley Dual-Plane: 544 hp @ 6,900 rpm
Holley Single-Plane: 552 hp @ 7,000 rpm
Largest Gain: 14 hp @ 7,200 rpm

The horsepower curves show a number of things, including the fact that the single-plane intake did indeed make more peak power than the dual-plane design, but not by much. Starting at 5,000 rpm, the single-plane pulled ahead, but the power difference was minimal. Run out to 7,200 rpm, the single-plane showed its worth by besting the dual-plane by 14 hp.

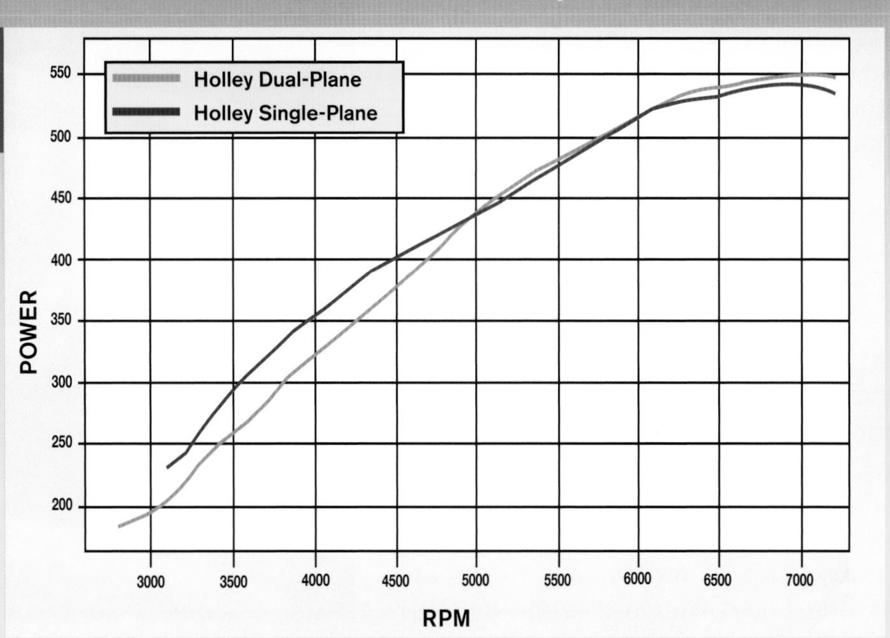

Holley Single- vs Dual-Plane Intake on an LS3 (Torque)

Holley Dual-Plane: 471 ft-lbs @ 4,300 rpm
Holley Single-Plane: 463 ft-lbs @ 5,200 rpm
Largest Gain: 54 ft-lbs @ 3,500 rpm

In terms of torque production, there really was no contest. Even though peak torque production differed by just 8 ft-lbs, the dual-plane offered gains that exceeded 50 ft-lbs down low. The additional torque offered by the dual-plane in the low- and mid-range is why it is usually chosen over the single-plane for most street applications. I even tested the dual-plane design under boost on a cathedral-port LS application with excellent results. If it's better naturally aspirated (NA), then it's better under boost!

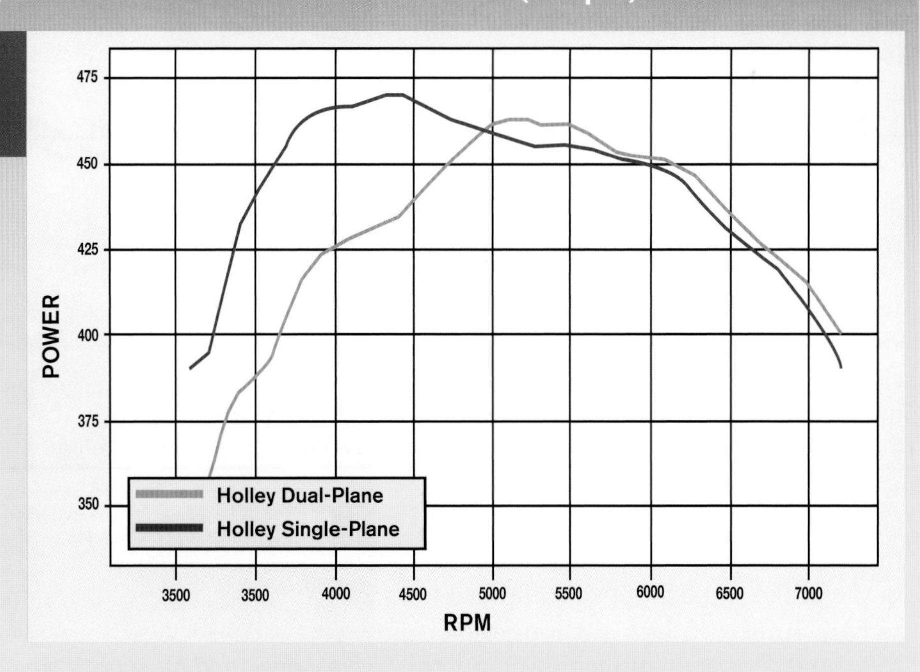

Test 2: FAST LSXR vs Mast Carbureted Single-Plane

This test offered a comparison between a long- and short-runner intake design. In fact, I used nearly the same 468-stroker test engine to compare the Mast single-plane intake to the FAST LSXR LS3 manifold. The 468 featured a Darton-sleeved block stuffed with a Lunati crank and rods teamed with JE forged pistons. Unlike the previous test, the 468 was topped with Mast black-Label LS3 heads.

The combination also included factory LS3 rockers, Comp hardened pushrods, and Kooks 1⅞-inch stainless headers. Also present was a Milodon oil pan and windage tray, Meziere electric water pump, and FAST 75-pound injectors. The finishing touch was, of course, the FAST LSXR LS3 intake manifold and 102-mm Big Mouth throttle body. Equipped with the FAST LSXR intake, the 468 produced 732 hp at 6,400 rpm and 665 ft-lbs of torque at 5,200 rpm.

After running the FAST intake, I replaced the EFI system (FAST XFI/XIM) with the Mast single-plane intake. The Mast intake featured a two-piece construction, which allowed them to fully CNC port the internals. This thing was a work of art; the kind you hate to install and get dirty.

The Mast intake was flanged to accept a 4500-series Holley carburetor. To feed the 468, I installed a Holley 1050 Ultra Dominator. The Mast intake was also designed to run in injected form, so I plugged the injector holes with a set of 19-pound Ford injectors. Equipped with the Mast intake, the power output of the 468 increased to 761 hp and 645 ft-lbs of torque. Not that the peak power rose, but the peak torque dropped compared to the FAST intake. In fact, the long-runner FAST intake offered more power up to 5,900 rpm, but the Mast single-plane pulled away up to 6,700 rpm.

The two-piece Mast LSX intake was designed for high-RPM high-horsepower LS3 applications. (The company also offers cathedral-port and LS7 versions.)

Likely designed for slightly smaller and milder applications, the FAST LSXR intake performed well on this 468-inch stroker. It is hard to argue with more than 730 hp from any manifold.

FAST LSXR vs Mast Carbureted Single-Plane (Horsepower)

FAST LSXR LS3: 732 hp @ 6,400 rpm
Mast Carbureted Single-Plane: 761 hp @ 6,500 rpm
Largest Gain: 47 hp @ 6,700 rpm

The high-RPM nature of the single-plane intake was evident in this curve. The CNC-ported Mast intake was a work of art and boy, did it pull hard on the top end. Unfortunately, all the top-end power came with a trade-off lower in the rev range; in this case, below 5,900 rpm.

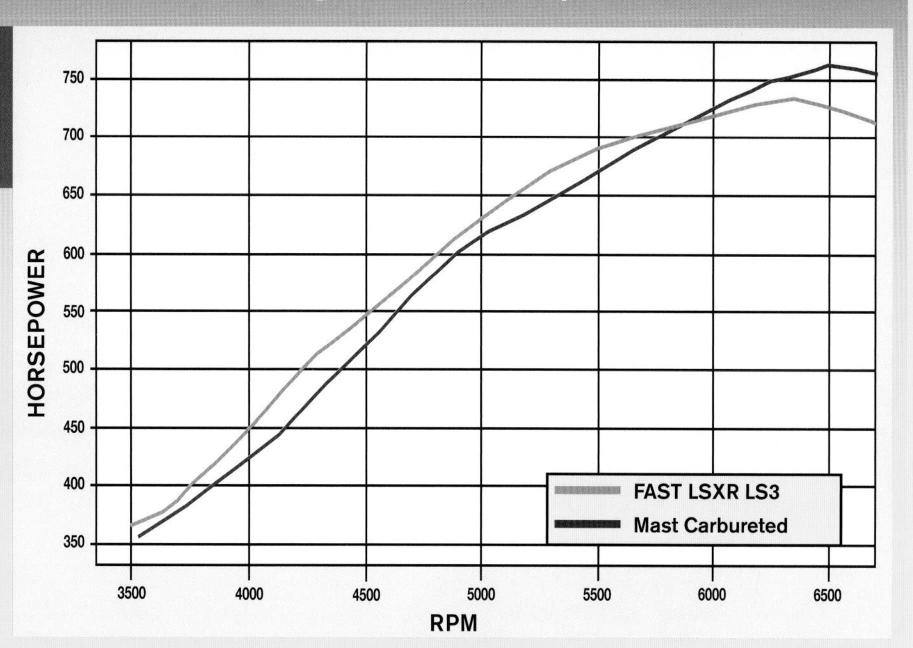

FAST LSXR vs Mast Carbureted Single-Plane (Torque)

FAST LSXR LS3: 614 ft-lbs @ 5,100 rpm
Mast Carbureted Single-Plane: 631 ft-lbs @ 5,300 rpm
Largest Gain: 37 ft-lbs @ 4,300 rpm

Much like the previous test run on the Holley Hi-Ram versus the stock LS3 intake, the long-runner FAST LSXR offered considerably more torque up to 5,850 rpm. The FAST offered torque gains as high as 37 ft-lbs lower in the rev range, but the high-RPM single-plane pulled away strong past 5,900 rpm.

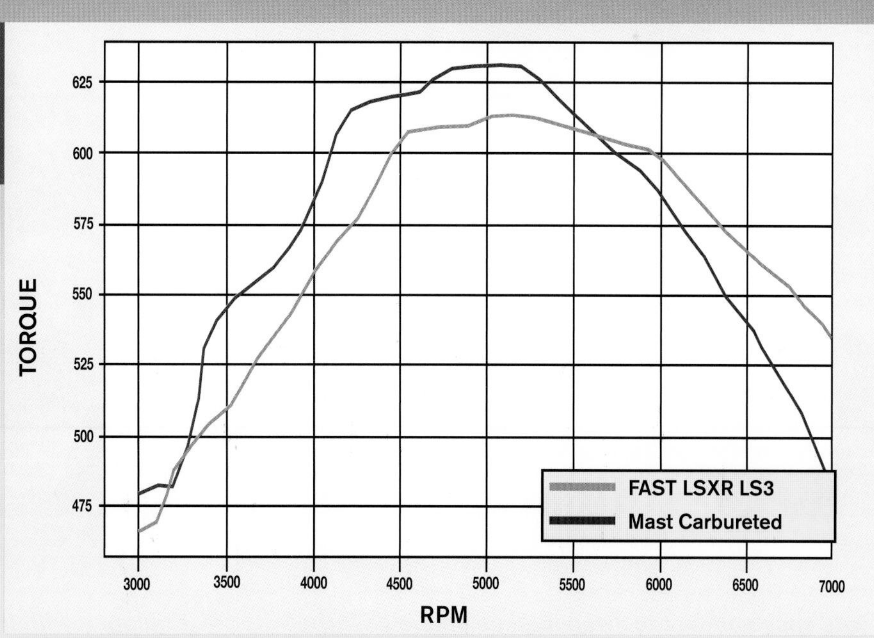

Test 3: Stock LS7 vs MSD Atomic for LS7, Modified LS7

This was one of those instances where an intake manifold swap did not trade low-speed torque for top-end power. In fact, this intake improved power through the entire curve, so you know it was the right choice for the engine combination.

Assembled by Cool Performance Machine, the 427 LS7 test engine was created by sleeving an LS3 block. The 4.130-inch bore received a complete Manley stroker assembly that included flat-top pistons, H-beam rods, and a Platinum-series (4.0-inch) stroker crank. Also included in the mix were Total Seal rings, a custom CMP cam (.644 lift and a 246/254-degree duration split), and an adjustable cam sprocket. In true LS7 fashion, the stroker featured an Aviad dry-sump oiling system. Feeding the over-bore LS7 was a set of CNC-ported CMP Brodix SI LS7 heads (395-cfm). Lucas 5W-20 synthetic oil, a Holley Dominator management system and Kooks long-tube headers were included in the mix. The build list also featured an ATI dampener, Meziere electric water pump, and FAST 102-mm throttle body.

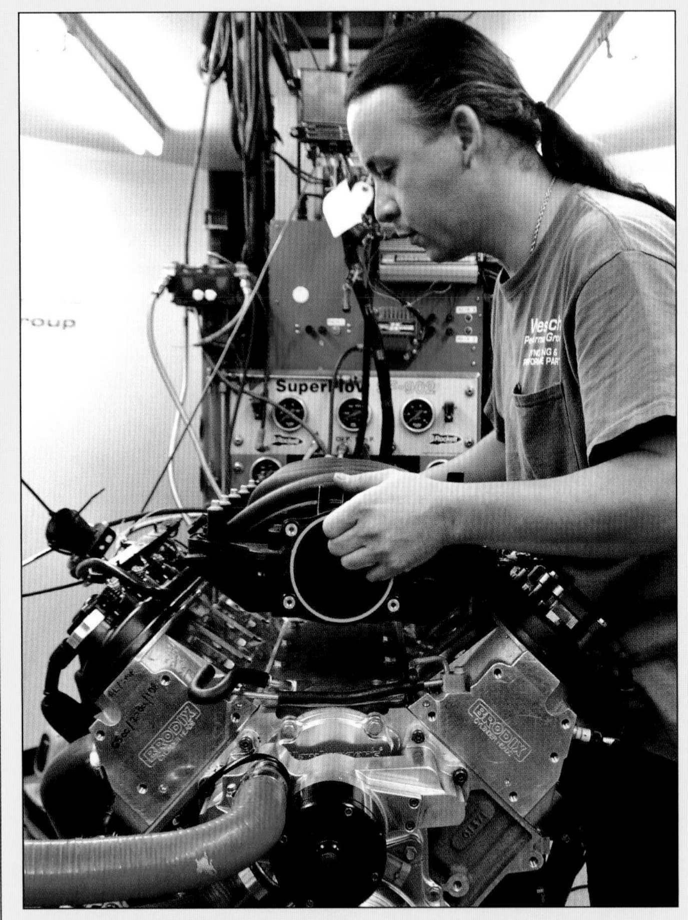

It was easy and fast to swap intakes on the LS7. The factory LS7 intake offered good power but nothing compared to the MSD Atomic.

This test involved running the stock LS7 composite intake against the MSD Atomic AirForce LS7 intake, which is also available for cathedral-port heads. Equipped with the stock LS7 intake, the modified LS7 produced 642 hp at 6,800 rpm and 554 ft-lbs of torque at 5,400 rpm. Torque production exceeded 540 ft-lbs for a 1,100-rpm spread (from 4,750 to 5,850 rpm). Since both intakes offered long(ish) runners, I was eager to see how well the MSD compared to the stock LS7.

After installation of the Atomic intake, I was immediately rewarded with both extra torque and horsepower. The peak numbers jumped to 684 hp and 586 ft-lbs of torque. The MSD intake offered nearly 20 ft-lbs below 3,500 rpm but as much as 40 ft-lbs elsewhere. The gains became serious after the tach passed 4,500 rpm. As much as I liked the extra 42 hp (peak-to-peak gain), I also liked the fact that the MSD improved the power output everywhere on this modified LS7.

Equipped with the stock LS7 intake, the CMP-modified LS7 produced 642 hp and 554 ft-lbs of torque. Both intakes were run with a FAST throttle body.

Stock LS7 vs MSD Atomic for LS7, Modified LS7 (Horsepower)

Stock LS7 Intake: 642 hp @ 6,800 rpm
MSD Atomic LS7 Intake: 684 hp @ 6,900 rpm
Largest Gain: 52 hp @ 6,000 rpm

The MSD Atomic intake started well and then became excellent as the tach zoomed past 4,500 rpm. Even down low, the Atomic offered improved power, but the intake really came alive after passing 4,500 rpm. Compared to the stock LS7 intake, the MSD increased the peak output by 42 hp but offered as much as 52 hp elsewhere in the curve.

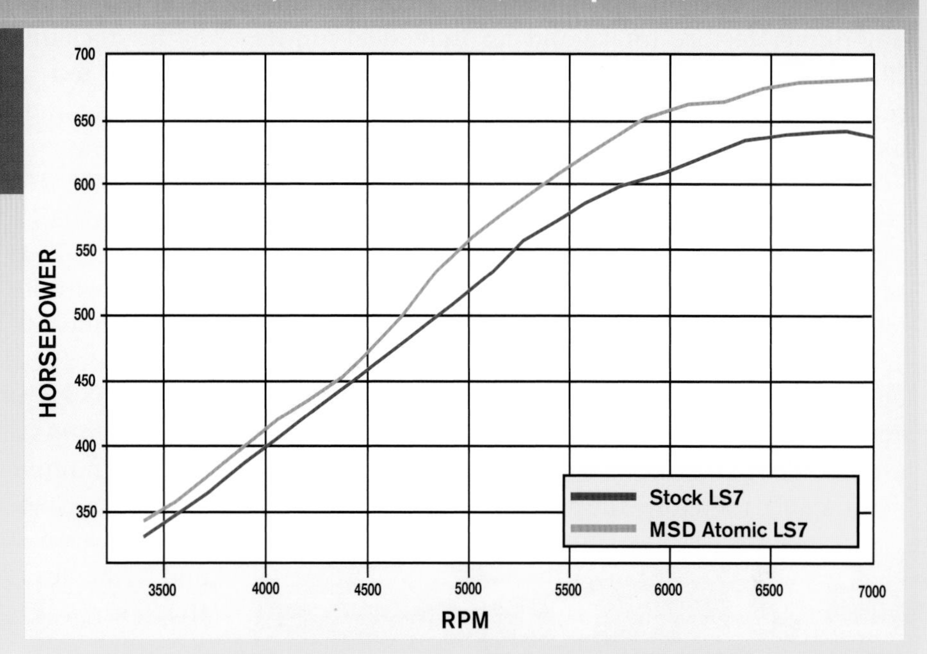

Stock LS7 vs MSD Atomic for LS7, Modified LS7 (Torque)

Stock LS7 Intake: 554 ft-lbs @ 5,400 rpm
MSD Atomic LS7 Intake: 586 ft-lbs @ 5,100 rpm
Largest Gain: 40 ft-lbs @ 5,000 rpm

As much as I love an extra 42 hp, I love extra torque through the entire rev range even more. As this chapter illustrates, top-end gains are often accompanied by losses in low-speed (and mid-range) torque. This was not the case on the Atomic intake test. The MSD AirForce intake improved torque production through the tested rev range and improved torque output by as much as 54 ft-lbs.

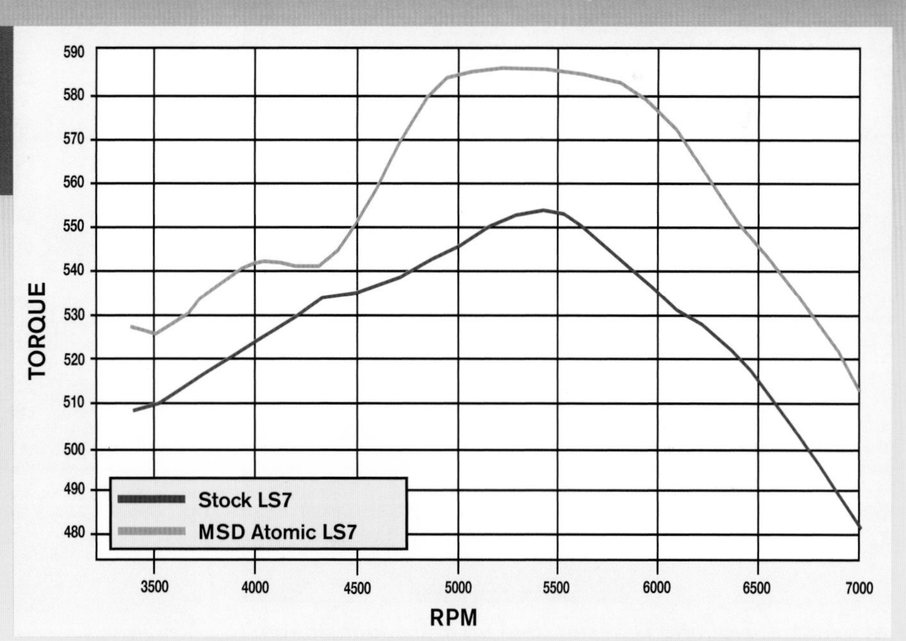

Test 4: Stock LS3 vs Speedmaster IR on a Modified LS3

More than just a change in runner length, this test involved a comparison between a conventional long-runner factory intake and an individual-runner (IR) intake from Speedmaster. The IR intake differed from the factory not only in the length of the runners (although they were different), but also in the lack of a common plenum. Each of the runners on the stock LS3 intake was connected to a common plenum fed by a single throttle body. By comparison, the Speedmaster intake featured no common plenum and eight individual throttle bodies, which measured 50 mm.

Some might be quick to point out that the IR system offered better airflow from the increased surface area of the eight throttle blades. However, the change in runner length, combined with the lack of a common plenum and, therefore, the absence of Helmholtz resonance are what really produced the power differences. (You know this because of the lack of vacuum present in the stock intake at wide-open throttle, or WOT.)

This test was a comparison between the stock LS3 intake and the Speedmaster IR manifold. The test engine was a crate LS3 upgraded with a Comp (PN 281LRRHR13) cam (.617/.624 lift split, 231/247 duration split, 113 LSA) and CNC-ported, TFS Gen X 255 heads. The LS3 crate engine from Gandrud Chevrolet was run with Kooks headers, a Holley HP management system, and Meziere electric water pump. Run on the dyno with the stock LS3 intake, the modified LS3 produced 575 hp at 6,500 rpm and 517 ft-lbs of torque at 5,500 rpm. As always, the long-runner, factory intake offered an impressive torque curve.

After installation of the Speedmaster IR intake, the peak power numbers jumped to 605 hp at 6,800 rpm and 533 ft-lbs of torque at 5,000 rpm. I liked the fact that both peak horsepower and torque were up, but also that the peak torque occurred lower in the rev range with the IR intake. With the exception of a small area near 4,500 rpm, the IR intake equaled or bettered the factory intake through the entire rev range.

Let's face it: Nothing is sexier than an individual-runner (IR) intake on an LS engine. This unit from Speedmaster offered impressive power gains to go along with its good looks.

The Speedmaster IR intake offered anodized fuel rails, full-radiused air horns, and individual ports to combine the MAP sensor readings.

Stock LS3 vs Speedmaster IR on a Modified LS3 (Horsepower)

Stock LS3 Intake: 575 hp @ 6,500 rpm
Speedmaster IR LS3 Intake: 605 hp @ 6,800 rpm
Largest Gain: 36 hp @ 6,800 rpm

With the exception of a slight dip in power near 4,500 rpm, the Speedmaster IR induction system improved the power output through the RPM range. There was a bump in power near 5,000 rpm and then a serious jump past 5,500 rpm. This intake would show even greater power gains on a larger or wilder application.

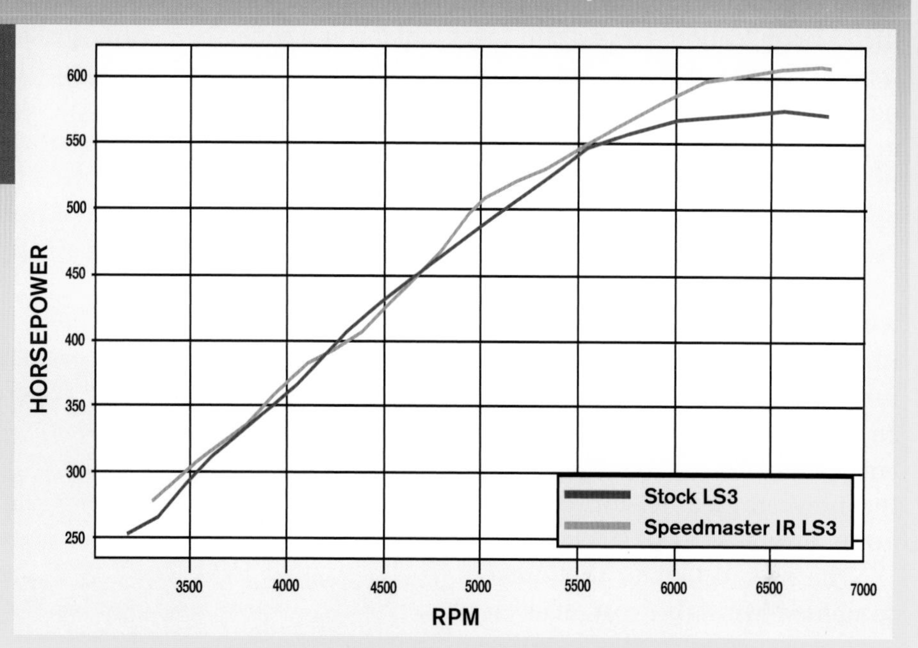

Stock LS3 vs Speedmaster IR on a Modified LS3 (Torque)

Stock LS3 Intake: 517 ft-lbs @ 5,500 rpm
Speedmaster IR LS3 Intake: 533 ft-lbs @ 5,000 rpm
Largest Gain: 22 ft-lbs @ 5,000 rpm

The torque curve shows the sign wave in power created by the IR intake from Speedmaster. The new induction system lost out in torque from 4,200 to 4,700 rpm, but bettered the stock LS3 intake everywhere else. A resonance wave at 5,000 rpm really bolstered torque production.

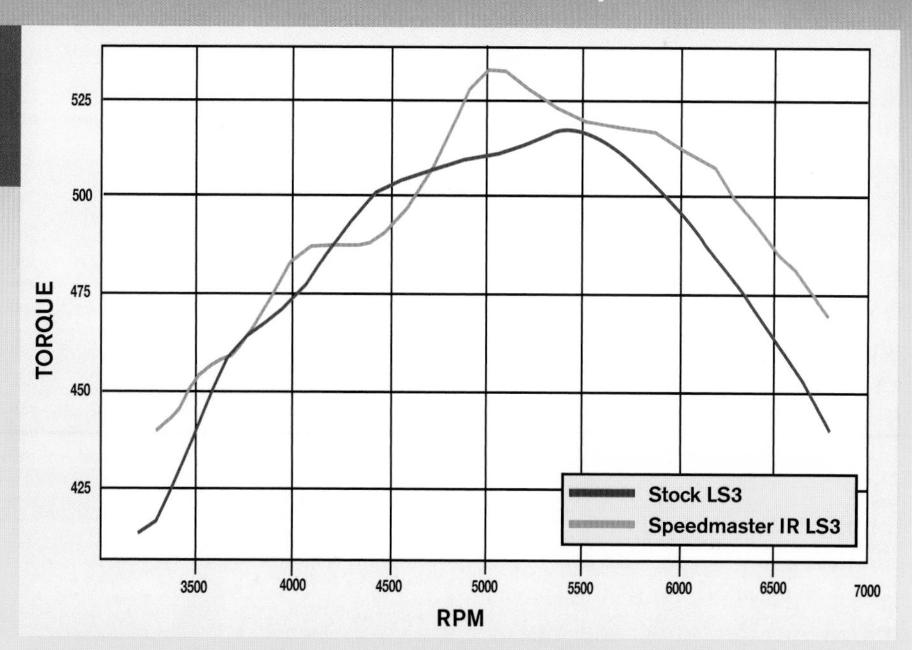

Test 5: Stock LS3 vs Speedmaster Fabricated Intake on a Mild LS3

A number of sources offer these fabricated intakes, but Speedmaster supplied this particular test piece for the LS3 application. Once again I was looking at a substantial change in runner length, to say nothing of plenum volume and throttle opening. Where the stock LS3 intake was designed to accept a 90-mm throttle body, the Speedmaster fabricated intake featured a 102-mm opening.

Various sources offer fabricated intake in different configurations, both with and without radiused air horns for the runners. If you plan to run one, make sure you select the one with air horns because the smooth air entry is worth 10 to 12 hp over the non-radiused version (at this power level).

The overall look of the intake combined with the cost make it a desirable commodity, but look over the dyno results before making your choice, especially for a mild street application. The intake certainly has its place, but like other short-runner intakes, there is a trade-off in low-speed (and mid-range) torque on all but the largest and wildest combos, which includes turbo and blower applications.

This test was run on an engine that clearly favored the long-runner, factory intake. The Gandrud LS3 crate engine was simply augmented with a mild 224 Crane cam (.624/.590 lift, 224/232 duration, 113 LSA), a set of long-tube 1¾-inch Quick Time Performance (QTP) headers with extensions, and mufflers. The test mule also featured a Holley 90-mm throttle body (stock intake), stock LS3 injectors (raised fuel pressure), and Holley HP management system. Run with the stock LS3 intake, the mild LS3 produced 538 hp at 6,300 rpm, and 504 ft-lbs of torque at 4,800 rpm.

After installation of the fabricated intake, the peak power jumped to 556 hp at (a higher) 6,800 rpm, but peak torque dropped to 485 ft-lbs of torque at 5,200 rpm. The stock intake offered improved power up to 6,050 rpm, where the fabricated began to pull away. This intake works much better on high-RPM or larger (and wilder) applications than on this mild LS3.

The factory LS3 intake is impressive, offering a good combination of power and torque production on most mild and modified LS3 applications.

Run with a mild cam, headers, and the stock intake, the LS3 produced peak numbers of 538 hp and 504 ft-lbs of torque.

Stock LS3 vs Speedmaster Fabricated Intake on a Mild LS3 (Horsepower)

Stock LS3 Intake: 538 hp @ 6,300 rpm
Speedmaster Fabricated Intake: 556 hp @ 6,800 rpm
Largest Gain: 32 hp @ 6,800 rpm

The Speedmaster fabricated intake offered impressive peak power gains, increasing the power output of the mild LS3 from 538 to 556 hp. It is important to note that the peak power occurred 500 rpm higher (from 6,300 to 6,800 rpm). This was a surefire indication of the high-RPM nature of the intake design.

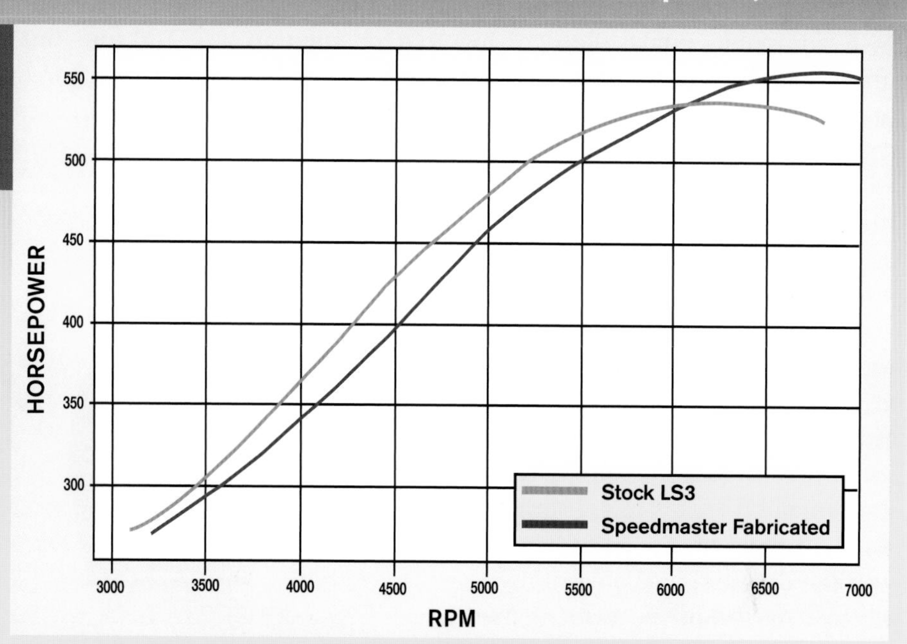

Stock LS3 vs Speedmaster Fabricated Intake on a Mild LS3 (Torque)

Stock LS3 Intake: 504 ft-lbs @ 4,800 rpm
Speedmaster Fabricated Intake: 485 ft-lbs @ 5,200 rpm
Largest Gain: 40 ft-lbs @ 4,500 rpm

The torque curve is even more telling because the long-runner, stock LS3 intake offered considerably more torque up to 6,000 rpm. As trick as they look, the short-runner intake designs are best left to high-RPM and/or large-displacement applications. The shorter runners lost as much as 40 ft-lbs of torque at 4,500 rpm.

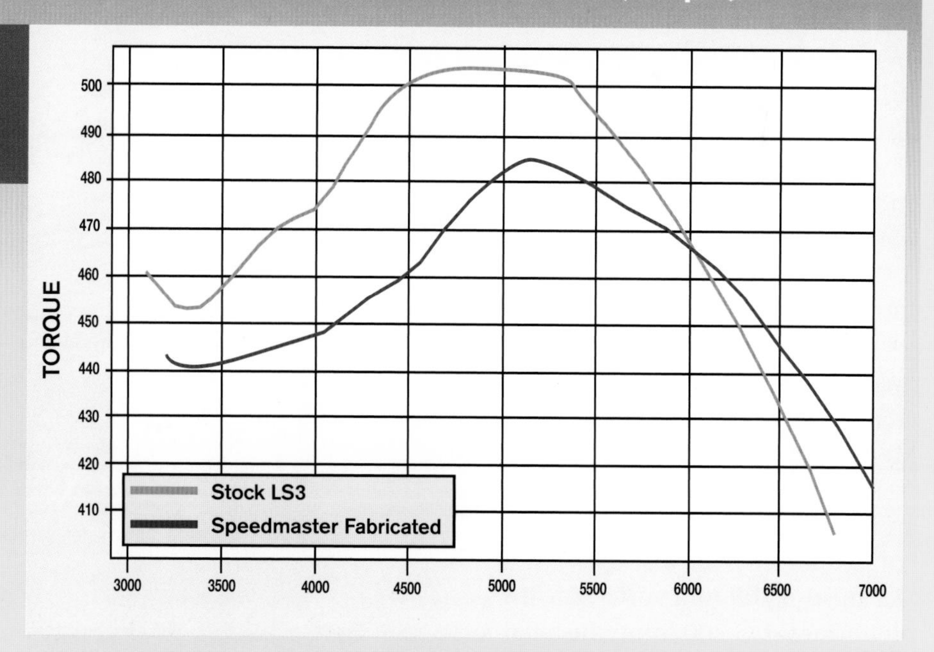

Test 6: FAST Adjustable LSXR Intake on a Mild LS3

Perhaps the best illustration of the effect of runner length comes from this test on the (then new) FAST LSXR adjustable intake. Recognizing that changes in runner length alter the effective operating range of the intake manifold, FAST designed the new intake to allow installation of different runner configurations. Using bolt-in runs, users can swap out runner lengths to tune the intake to different engine combinations and/or applications (think street/strip).

One thing I also tested but did not show in this book was to combine different lengths (four short and four long or four medium), not unlike a single-plane, carbureted intake. For this test, I simply ran the intake on a cam-only LS3 with the different available runner lengths. As should be evident by now from the results of the previous tests in this chapter, shorter runner lengths increased peak power but traded low-speed torque.

The test engine was a crate LS3 from Gandrud Chevrolet augmented with a cam from Brian Tooley Racing (BTR). The grind featured a .615/.595 lift split, a 229/244-degree duration split, and 113-degree (+4) LSA. The cam was combined with a dual-spring upgrade to replace the factory LS3 springs. Other components used on the test engine included a FAST management system and injectors, Hooker long-tube headers, and a Meziere electric water pump.

Running 5 quarts of Lucas oil and the longest of the three runner configurations, the LS3 produced 562 hp at 6,400 rpm and 512 ft-lbs of torque at 5,200 rpm. Installation of the medium-length runners increased peak power to 568 hp at 6,800 rpm, but torque dropped to 490 ft-lbs at 5,300 rpm. The final test involved installation of the shortest runners that resulted in 577 hp at 7,100 rpm, but torque dropped further still to 478 ft-lbs at 5,300 rpm. Each successive decrease in runner length resulted in increase peak power but a drop in torque (in this case, below 6,500 rpm).

The adjustable intake offered three bolt-in runner lengths to dial in the power curve to a specific combination.

The LSXR adjustable intake looked just like the original, but under the lid was a surprise.

FAST Adjustable LSXR Intake on a Mild LS3 (Horsepower)

FAST LSXR Long Runner: 562 hp @ 6,400 rpm
FAST LSXR Medium Runner: 568 hp @ 7,000 rpm
FAST LSXR Short Runner: 577 hp @ 7,100 rpm
Largest Gain: 20 hp @ 6,700 rpm

Looking at the numbers, you might be tempted to pick the intake combination that offered the highest peak power. Unfortunately, we do not live by peak power alone. While the middle and shorter runner offered higher peak power numbers, there was a trade-off in power elsewhere in the curve. Only above 6,500 rpm and below 3,400 rpm did the middle or shorter runners offer more power.

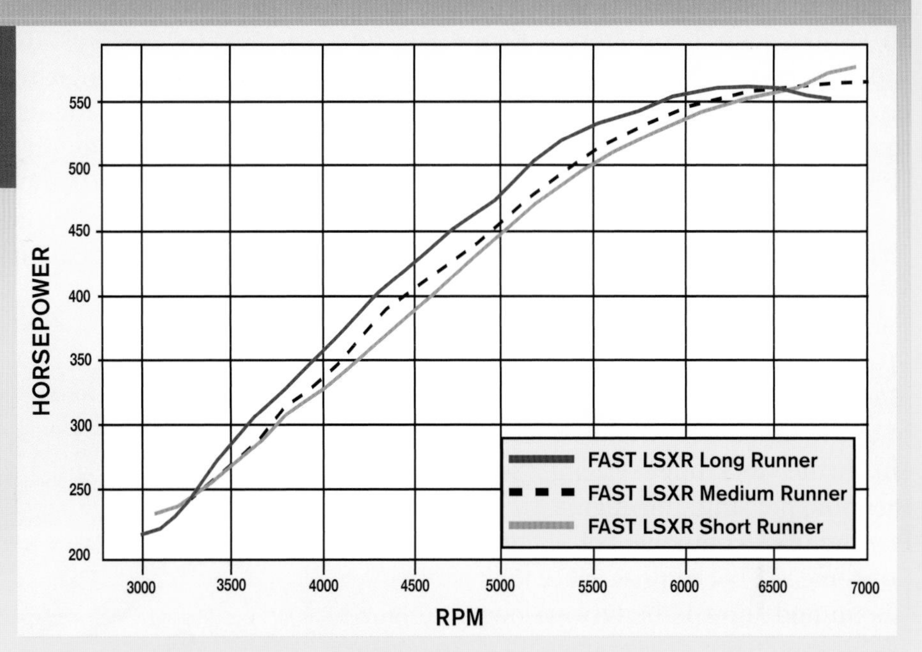

FAST Adjustable LSXR Intake on a Mild LS3 (Torque)

FAST LSXR Long Runner: 512 ft-lbs @ 5,200 rpm
FAST LSXR Medium Runner: 490 ft-lbs @ 5,300 rpm
FAST LSXR Short Runner: 478 ft-lbs @ 5,300 rpm
Largest Gain: 45 ft-lbs 4,400 rpm

The results of this test on the mild (cam-only) LS3 clearly show the effects of changes in runner length. The trick LSXR adjustable intake allowed me to show the drop (or increase) in torque at lower engine speeds offered by shortening or lengthening the runners. The shortest runner offered the lowest peak torque, followed by the middle runners. The longest runners offered the highest peak (and average) torque production.

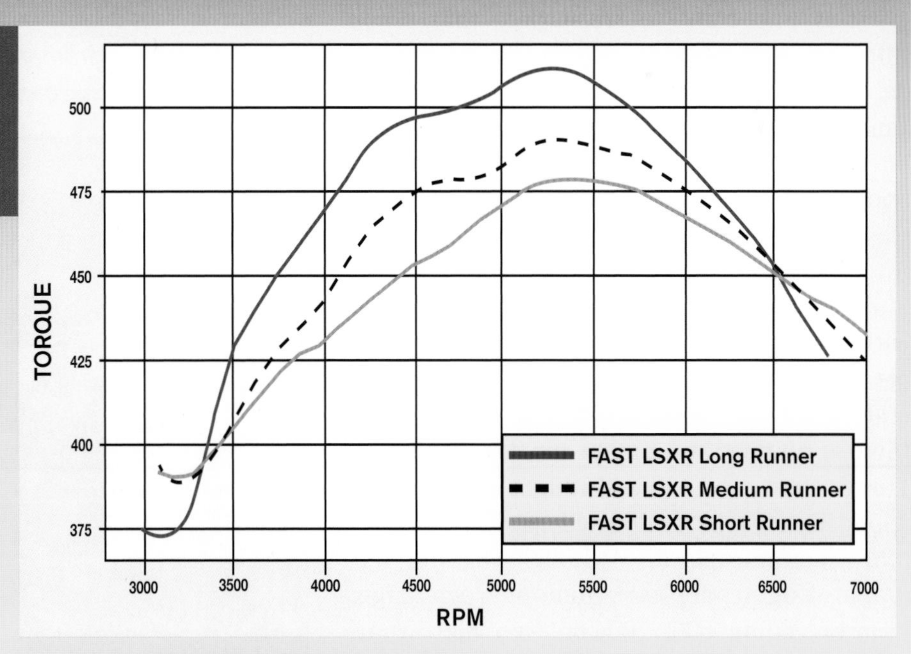

Test 7: Stock vs Kenne Bell 102-mm Throttle Body on a KB SC LS3

About the only thing better than a modified LS engine is a supercharged one. Nothing adds zing to an LS like some boost from a Kenne Bell twin-screw supercharger. While twin-screw kits are efficient and powerful, boost is only the beginning. The reality is that superchargers are only as good as their induction system.

Nothing chokes off the power potential of a supercharger faster than a restrictive throttle body or associated inlet components. Knowing this, the question is, How much power is a throttle body upgrade really worth? As an airflow device, the modified power output determines the amount of power hindered by the flow restriction inherent in the stock inlet system. This means that the more powerful the engine, the more restrictive the stock components become. This should not come as a big surprise, since the factory inlet system and throttle body were never designed for the elevated power levels offered by a Kenne Bell supercharger. The inlet system that General Motors designed to support 425 hp has no business on a supercharged engine making 600 or more horsepower.

It is important to stress here that power gains offered by the throttle body are entirely dependent on the engine combination. As a simple airflow device, the higher the power output of the test engine, the larger the throttle body required.

As an example, installation of a larger throttle body capable of supporting 1,000 hp is of little use on a 425-hp engine equipped with an (already oversized) throttle body capable of supporting 750 hp. The 750-hp throttle body is already oversized for the application, so there is no need to upgrade on the NA engine. Things change on (draw-through) supercharged applications, where elevated power levels are more commonplace. Although 600-hp NA Camaro engines are less common, supercharged LS3s exceeding 600, 700, or even 800 hp are everywhere.

This round of testing on a Kenne Bell supercharged (2010 LS3) Camaro illustrated that a throttle body upgrade on a 600-hp application (9.3 psi on stock engine) was worth 8 hp. Performing the same test at 13 psi (678 hp) was worth 26 hp (up to 702 hp) and an amazing 34 hp at 17 psi (from 755 hp to 789 hp). The higher the boost (and power) run on the test engine, the greater the losses associated with a restrictive throttle body. It is important to note that testing the same throttle body upgrade on the NA LS3 was worth 0 extra hp.

Airflow increases dramatically when you install a Kenne Bell twin-screw supercharger.

This round of testing was performed on a 2010 LS3 Camaro equipped with a 2.8L, twin-screw supercharger kit from Kenne Bell.

Stock vs Kenne Bell 102-mm Throttle Body (13psi) (Horsepower)

Stock 90-mm LS3 TB: 678 hp @ 4,300 rpm
KB 102-mm TB: 702 hp @ 4,400 rpm
Largest Gain: 26 hp @ 4,700 rpm

A throttle-body upgrade offers power gains that are in relation to the power output of the test engine. Tested on the stock NA, the larger throttle body was worth nothing. Tested at 13 psi, the upgrade was worth 26 hp.

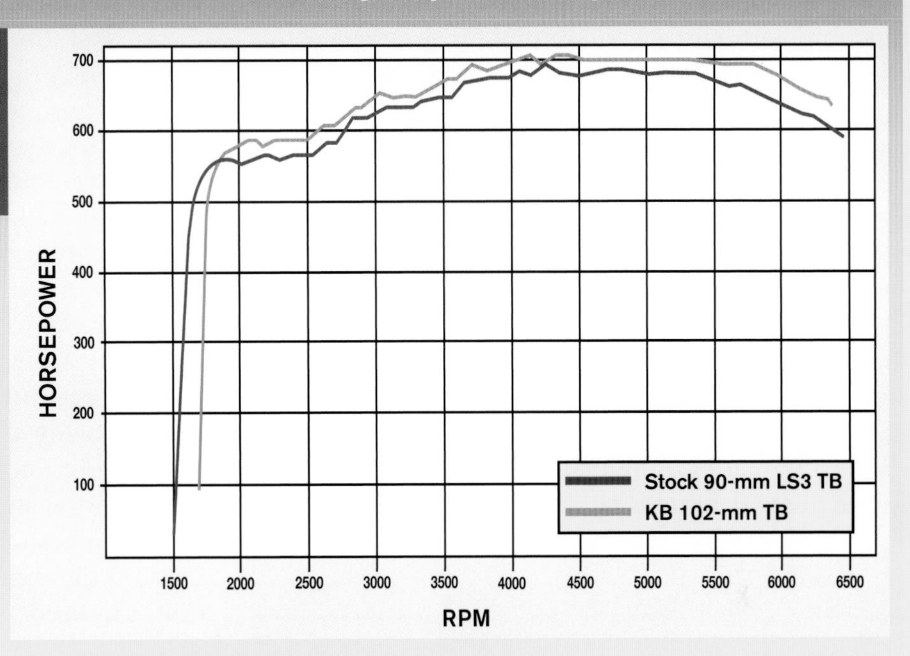

Stock vs Kenne Bell 102-MM Throttle Body (17psi) (Horsepower)

Stock 90-mm LS3 TB: 755 hp @ 6,300 rpm
KB 102-mm TB: 789 hp @ 6,400 rpm
Largest Gain: 34 hp 6,400 rpm

Running the throttle-body test at a higher boost level (17.3) resulted in a significant rise in power. Run at this elevated power level, the throttle-body upgrade was worth 34 hp and increased boost by 1 psi.

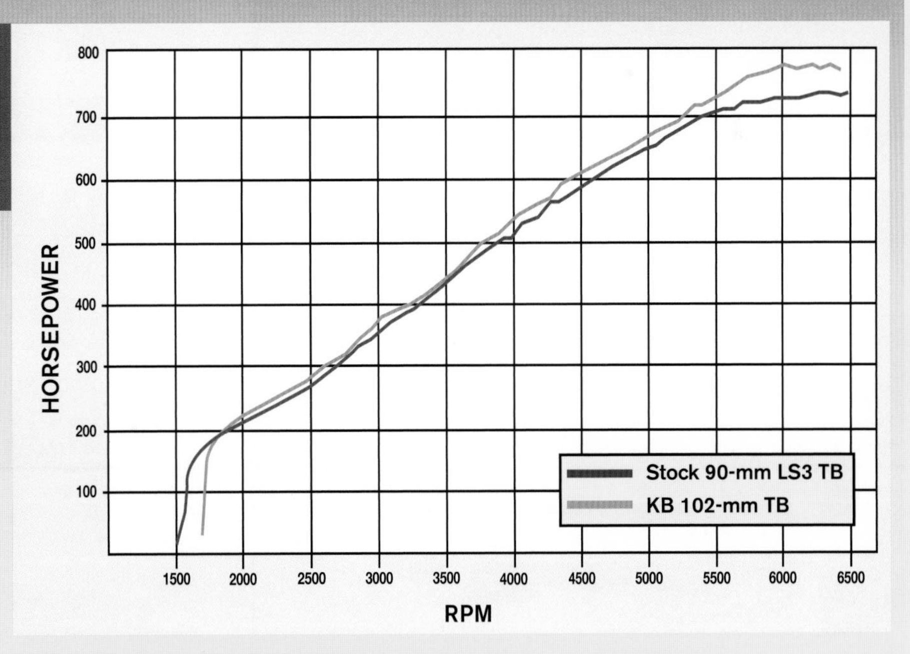

Test 8: Custom Dual-Plenum Adjustable-Runner Intake on a 468 Stroker

Long before the introduction of the FAST Adjustable LS3 or Edelbrock Cross-Ram intake, enthusiasts were tinkering with custom intake designs. I designed this adjustable intake for LS3-headed applications in 2008 to illustrate changes in the power curve. In addition to the dual-plenum design (with removable plenum connection), I was able to quickly adjust the runner lengths to optimize power production at different engine speeds.

The runner length acts as a tuning device to tailor the shape of the power curve. Longer runners optimize power production lower in the rev range than shorter runners. The downside to any given length is that there are trade-offs at the other end of the rev range. The additional low- and mid-range torque offered by longer runners is offset by a loss in high-RPM power. The opposite is true of short runners because they give up low- and mid-range torque for optimization at high RPM. The idea is to tune the combination for the desired use.

The test engine was a 468 stroker that was made possible by combining an LS6 block with Darton sleeves to allow for a 4.185-inch bore.

To test the custom intake, I needed an engine capable of using the massive flow capability of the 2.25-inch runners. In short, I needed something more than either a stock LS3 or LS7. Knowing this, I assembled a big-bore, LS stroker engine by combining a Darton-sleeved LS6 block with a Lunati stroker crank and K1 connecting rods. The Darton MID sleeve system provided the necessary room to allow me to bore the block out to 4.185 inches.

I then combined the big bore with a 4.25-inch Lunati forged-steel stroker crank. The result was a stroker displacing a massive 468 ci, or more than enough to properly test the merits of the custom intake system. The 468 also featured a static compression ratio of 12.25:1, a healthy 305LRR HR15 Comp cam (.624 lift, a 255/271 duration split, and 115 LSA), and Speedmaster CNC LS3 heads.

Run with the short (7.25-inch) runners, the 468 produced 723 hp and 620 ft-lbs of torque, but these numbers changed to 704 hp and 638 ft-lbs with 10.5-inch runners, then to 688 hp and 648 ft-lbs with the longest 16.5-inch runners tested. As length increased so did torque production, but the peak power fell off. Such is the trade-off inherent in runner length.

The sleeved block was then treated to a Lunati 4.25-inch stroker crank and K1 rods, along with a set of Wiseco forged pistons.

Custom Dual-Plenum Adjustable-Runner Intake on a 468 Stroker (Horsepower)

16.5-inch Runners: 688 hp @ 6,500 rpm
10.5-inch Runners: 704 hp @ 6,400 rpm
7.25-inch Runners: 723 hp @ 6,900 rpm
Largest Gain: 36 hp @ 6,800 rpm

As I saw with the FAST Adjustable LS3 intake, adjusting the runner length on this custom, dual-plenum intake on the 468 stroker had a similar effect on power production. Shorter runners push power production higher in the rev range; longer runners optimize power at lower engine speeds. The 16½-inch runners offered the most power up to 5,400 rpm, but lost out to the shorter 10½ and 7¼-inch runners thereafter.

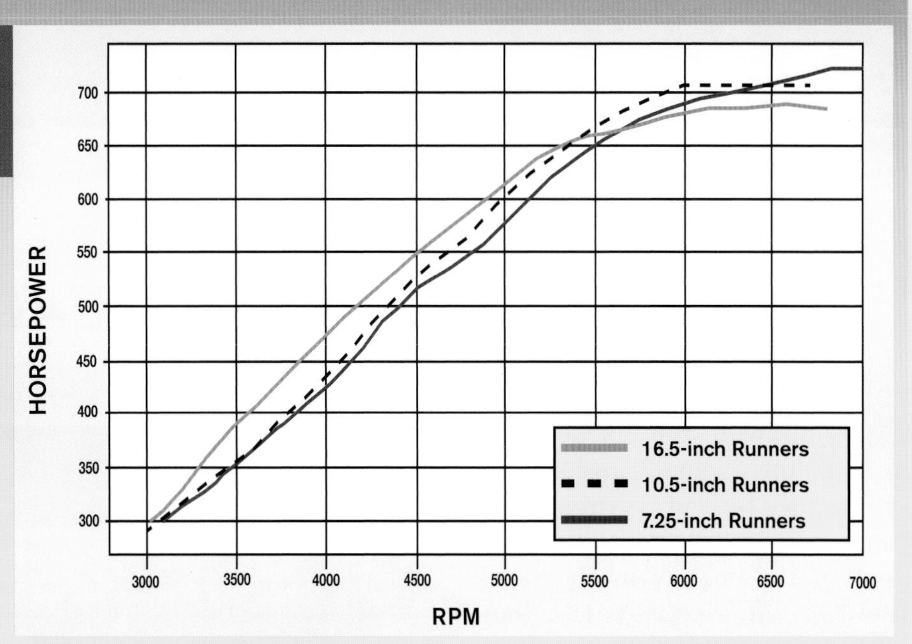

Custom Dual-Plenum Adjustable-Runner Intake on a 468 Stroker (Torque)

16.5-inch Runners: 648 ft-lbs @ 5,100 rpm
10.5-inch Runners: 638 ft-lbs @ 5,400 rpm
7.25-inch Runners: 620 ft-lbs @ 5,400 rpm
Largest Gain: 62 ft-lbs @ 3,900 rpm

The 16½-inch runners were the clear winner in torque production up to the crossover point of 5,400 rpm. The torque gains were as high as 62 ft-lbs at 3,600 rpm over the shortest runner length. The 10⅕-inch runners offered more torque than the 7¼-inch runners all the way up to 6,500 rpm. Only above that did the short-runner combination excel.

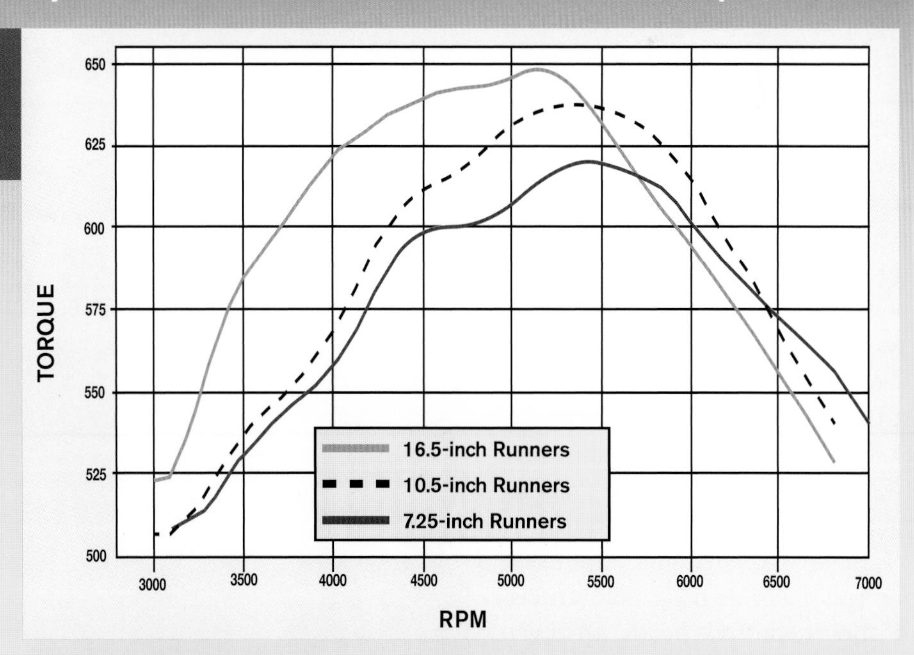

Cylinder Heads

Working with the intake and camshaft, the cylinder heads are part of the trio of performance components that dictate the power output of the engine. In the case of the LS3 and (especially) LS7, the factory heads offer exceptional airflow. Unlike cathedral-port heads (706, 317, 243, etc.), it is difficult to improve upon the power output of the already impressive factory heads.

I have seen power gains eclipsing 70 hp when upgrading cathedral-port heads (on a 408 stroker), but the gains were nearly half that (or less) when replacing factory LS3 heads (on a larger 468 stroker). The reason for this is not that the aftermarket doesn't know how to produce a good LS3 head, but that the factory LS3 heads already flow enough to support such high power levels. A stock LS3 head flows near 318 cfm. This compares to the very best cathedral-port head (317 or 243) that flows 244 cfm. The difference between an LS3 and the best factory cathedral-port is more than 70 cfm.

I have exceeded 690 hp using stock LS3 castings on a 468 stroker, and the stock LS7 heads are even more impressive. Run on a 495-inch stroker, the stock LS7 heads produced 773 hp. The impressive head flow

Bolting on the right set of LS3 or LS7 cylinder heads can yield impressive power gains.

Factory LS3 and LS7 heads offer exceptional flow and power potential, but CNC-ported, aftermarket heads offer even more.

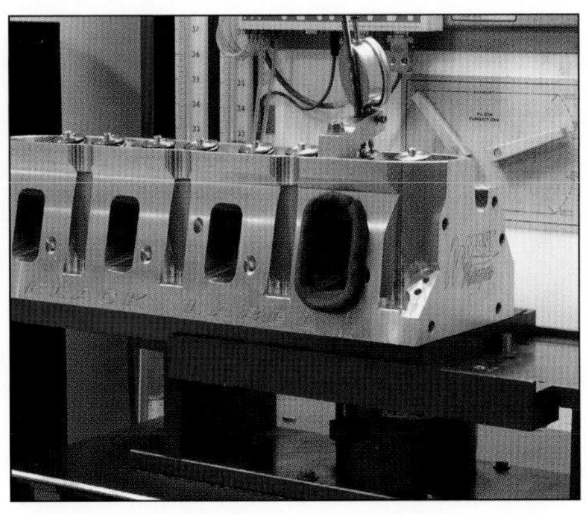

Using a flow bench is one way to determine the power "potential" of a set of cylinder heads, but the only way to know for sure is to run them on a dyno.

For the ultimate in valve control and RPM potential, a dual-spring upgrade is the way to go on a high-performance LS3 or LS7.

offered by the stock heads is both a blessing and a curse. On the plus side, they offer impressive power right out of the box, but just don't expect huge power gains when upgrading the heads on your LS3 or LS7.

To understand the reason for this, you need to first understand the correlation between airflow and power potential. There is, of course, an equation to calculate the "potential" horsepower offered by cylinder heads using the airflow data. This formula is:

$$HP = .257 \times \text{airflow} \times \text{number of cylinders}$$

Using an LS3 (with 318 cfm) as an example, the formula indicates (.257 x 234 cfm x 8) that the stock heads will support 653 hp (though have made more than the formula suggests). Were you to upgrade the heads on a stock or mild LS3 (making less than 560 hp), the gains offered by the head swap might be minimal because the stock heads already flow more than enough to support the current power level. Test 1 in this chapter illustrates what happens when you add cylinder head flow to a mild combination.

LS3 and LS7 heads offered by the aftermarket are usually purchased based on flow numbers. The problem with purchasing cylinder heads based on airflow is that the airflow numbers represent only a potential power output. As in the example in Test 1, just because you have 800-hp head flow doesn't mean your combination is in a position to take full advantage of the available flow. This is especially the case in LS7 applications, where aftermarket head flow can exceed 400 cfm (or more). It took a 495-inch super stroker (see Chapter 8) to tax the flow limits of the best LS7 heads, and the stock LS7 heads produced 773 hp.

Even the best heads were only up by 25 hp or so, the gains offered by ported LS7 heads would be even less. On a stock LS7, there may be no gain at all. This is especially the case if the maximum flow rate given for the heads you plan to purchase exceed the lift of the cam you plan to run. Big flow at .700, .750, or .800 lift is useless if you plan to use a .600-lift cam. Besides, you should be more concerned with the mid-lift flow numbers because the valve spends much more of its time sweeping through the mid-lift (opening and closing) than it does at peak lift.

The great thing about the LS engine family is the interchangeability. Cylinder heads from an LS7 physically bolt onto an original LS1, but the small bore size does not allow them to actually run without valve interference. This interchange allows the later LS3 heads to serve as inexpensive upgrades to the earlier cathedral-port engines. The most popular upgrade is to replace the stock 317 cathedral-port heads on a 6.0 truck (LQ4 or LQ9) or LS2 (243 castings) with the rectangular-port LS3 heads. The large valves in the LS3 heads require the 4.0-inch bore of the 6.0 (and do not work on smaller 4.8, 5.3, or 5.7 blocks), but the results are impressive.

The LS3 head upgrade also requires the corresponding offset (intake) rockers and intake manifold, but the stock LS3 heads offer an additional 70 cfm per runner. Tested on a 408 stroker, the LS3 head upgrade was worth almost 40 hp over the stock 317 truck heads. The smaller cathedral-port 317 heads offered more power up to 4,000 rpm, but the LS3 heads pulled away up to 6,500 rpm.

Test 1: Stock LS3 vs Chevy Performance CNC L92 on a Stock LS3

Several tests in this book were designed to illustrate what happens when you install the right part on the wrong application. Unlike factory cathedral-port applications, LS3 engines were blessed with high-flow cylinder heads. In terms of head flow, there was a substantial step up from the cathedral-port LS6/LS2 heads to the rectangular-port LS3 heads. That is why adding LS3 heads to a 6.0 is such a popular swap.

As you learn in this chapter, factory heads can support nearly 700 hp on the right application, but that doesn't mean ported heads don't offer any power. Just don't expect the huge gains normally seen with cathedral-port head testing; the stock LS3 heads flow nearly 315 cfm. This test shows that, especially on a stock application, cylinder head flow was not the limiting factor in terms of performance.

This test was run on the LS3 crate engine from Gandrud Chevrolet in near-stock trim. The engine was equipped with a set of long-tube headers, manual FAST throttle body, and Holley HP management system. Everything else on the engine was left stock, including the camshaft, displacement, and compression ratio. This test was run to illustrate what happens when you increase the head flow on an engine that already has enough cylinder head.

Run with the stock LS3 heads, the stock LS3 produced 493 hp at 5,700 rpm and 484 ft-lbs of torque at 4,800 rpm. I then installed a set of CNC-ported L92 heads from GM Performance (supplied by Gandrud) that flowed nearly 350 cfm (up from 315 cfm). While the additional head flow could support well over 700 hp, run on this stock application, the ported heads improved the power output to only 503 hp and 497 ft-lbs of torque. Does this mean the CNC-ported heads don't work? Hardly. See the gains offered in Test 3.

The test engine was an LS3 crate engine supplied by Gandrud Chevrolet. It was run in stock trim with no changes to the cam, compression, or induction system.

For dyno use, I installed a set of long-tube headers, FAST throttle body, and XFI management system. The crate engine produced 493 hp and 484 ft-lbs of torque with the stock heads.

Stock LS3 vs Chevy Performance CNC L92 on a Stock LS3 (Horsepower)

Stock LS3 Heads: 493 hp @ 5,700 rpm
Chevy Performance CNC L92 Heads: 503 hp @ 5,900 rpm
Largest Gain: 13 hp @ 6,400 rpm

Replacing the stock LS3 heads with a set of CNC-ported L92 from Chevy Performance netted only minor gains on this otherwise stock LS3 crate engine. The CNC-ported heads flowed considerably more than the stocks, but the mild stock engine simply didn't need any more flow to support the existing power level. Test 3 illustrates the amount of power that the extra airflow can be worth on the right application.

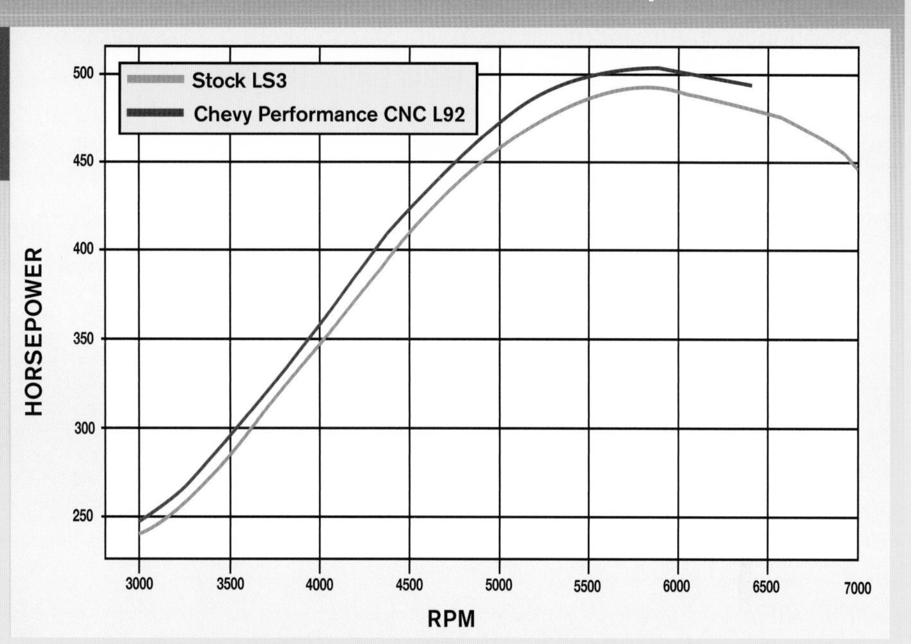

Stock LS3 vs Chevy Performance CNC L92 on a Stock LS3 (Torque)

Stock LS3 Heads: 484 ft-lbs @ 4,800 rpm
Chevy Performance CNC L92 Heads: 497 ft-lbs @ 4,600 rpm
Largest Gain: 16 ft-lbs @ 4,400 rpm

The torque curve tells the same story because the mild LS3 simply couldn't use the additional airflow. Because the stock LS3 heads support well over 600 hp, they were more than enough for this stock crate engine. The head swap would be worth much more than 16 ft-lbs on the right application.

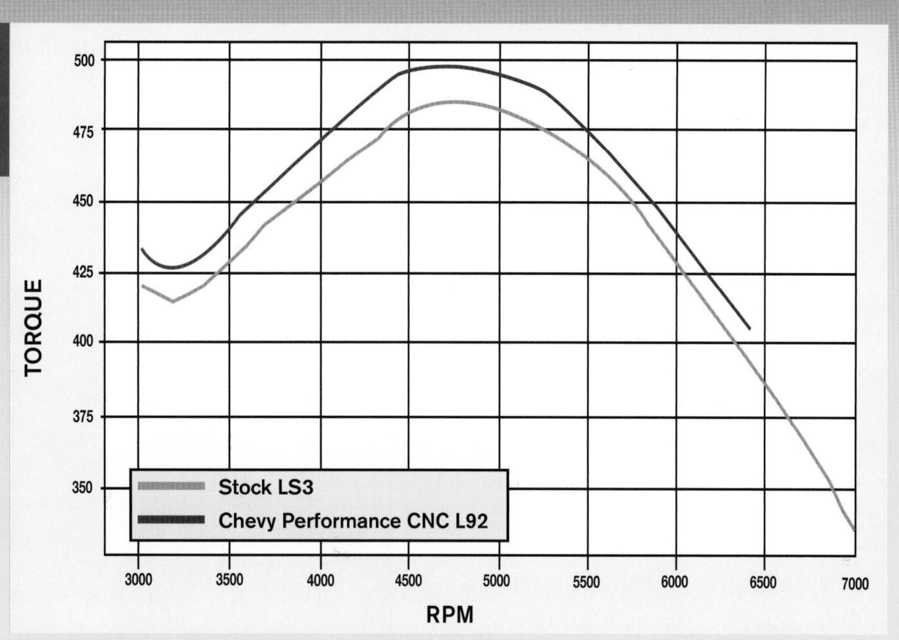

Test 2: Stock LS3 vs AFR 245 on a 408 Stroker

Since both the AFR 245 cathedral-port and factory LS3 heads used for this test flowed so well, I made sure I had a solid test engine for the comparison. Starting with a 6.0 block, the 408 was assembled using components from Speedmaster and Wiseco. The stroker assembly included a 4.0-inch forged-steel crank combined with 6.125-inch 4340 forged steel connecting rods and 10-cc dished pistons.

The forged pistons from Wiseco featured valve reliefs to allow for the hydraulic roller (PN 289LRR HR14) cam (.624 lift, a 239/255 duration split, and 114 LSA) from Comp Cams. Designed for a rectangular-port head application (which may not have favored the cathedral-port heads), the cam was combined with a set of standard-travel lifters and hardened pushrods from Comp Cams.

The .030-over 408 stroker short-block also featured a new timing chain and oil pump from Speed Pro, a set of ARP head studs and MLS head gaskets from Fel Pro, and a Moroso oil pan and windage tray.

Comparing the rectangular-port LS3 and the cathedral-port AFR heads also required an intake swap (to match the respective head ports). To keep the test as even as possible, I selected FAST LSXR intakes for both applications. Both were also run with the same 102-mm Big Mouth throttle body. As always, both heads were run with the same air/fuel ratio and timing values.

Equipped with the stock LS3 heads, the 408 produced 581 hp and 543 ft-lbs of torque. After installation of the AFR 245 heads, the peak numbers jumped to 604 hp and 665 ft-lbs of torque. It must be pointed out that the head swap also included a change in static compression ratio because the chamber volume on the two heads differed by 5 cc (64 cc vs 69 cc). This meant that in addition to the increased airflow offered by the AFR heads (349 cfm vs 316 cfm), they also increased the static compression ratio by .5 points. With the exception of a short, 250-rpm range (from 4,100 to 4,250 rpm), the AFR heads improved the power output from 3,000 to 6,700 rpm. Having more peak power is good, but having extra power everywhere is even better.

The 408 stroker test engine started as an iron 6.0 block but was stuffed with a Speedmaster forged crank and rods and Probe Racing dished pistons. Note the ARP head studs and Fel Pro MLS head gaskets.

Equipped with the stock LS3 heads and FAST LSXR LS3 intake, the 408 stroker produced 581 hp and 543 ft-lbs of torque.

Stock LS3 vs AFR 245 on a 408 Stroker (Horsepower)

Stock LS3 Heads: 581 hp @ 6,200 rpm
AFR 245 Heads: 604 hp @ 6,500 rpm
Largest Gain: 27 hp @ 5,700 rpm

Replacing the stock LS3 heads on the 408 stroker with a set of cathedral-port AFR 245 heads netted impressive results. The combination of increased flow and compression increased the power output from 581 to 604 hp, but the head swap increased power through most of the curve.

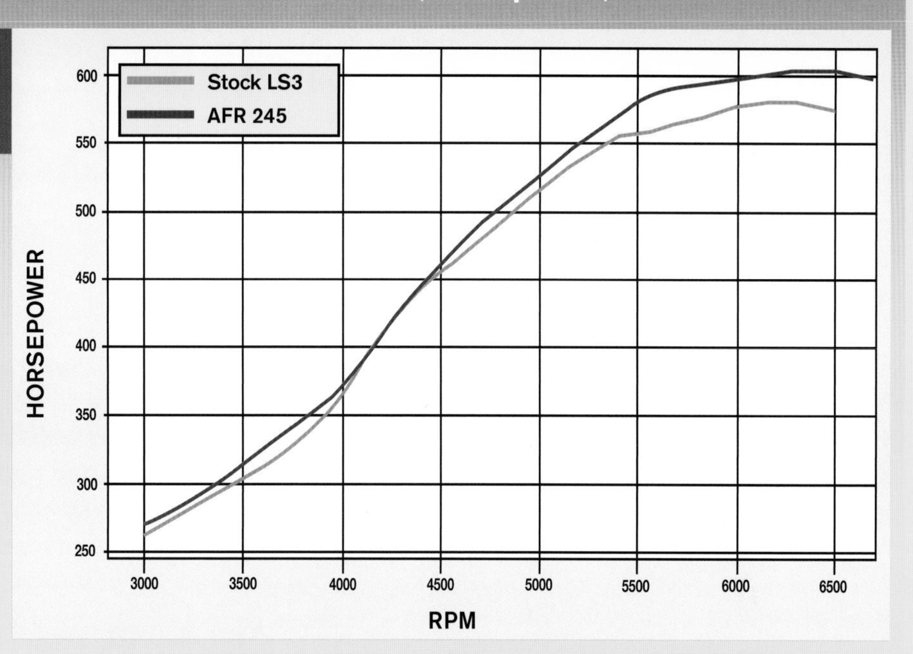

Stock LS3 vs AFR 245 on a 408 Stroker (Torque)

Stock LS3 Heads: 543 ft-lbs @ 5,100 rpm
AFR 245 Heads: 556 ft-lbs @ 5,400 rpm
Largest Gain: 24 ft-lbs @ 5,600 rpm

The cathedral-port AFR heads improved torque production down low by as much as 18 ft-lbs, but the gains were even more significant above 4,500 rpm. With the exception of a 100-rpm spread (from 4,100 to 4,200 rpm), the head swap improved torque production through the tested rev range.

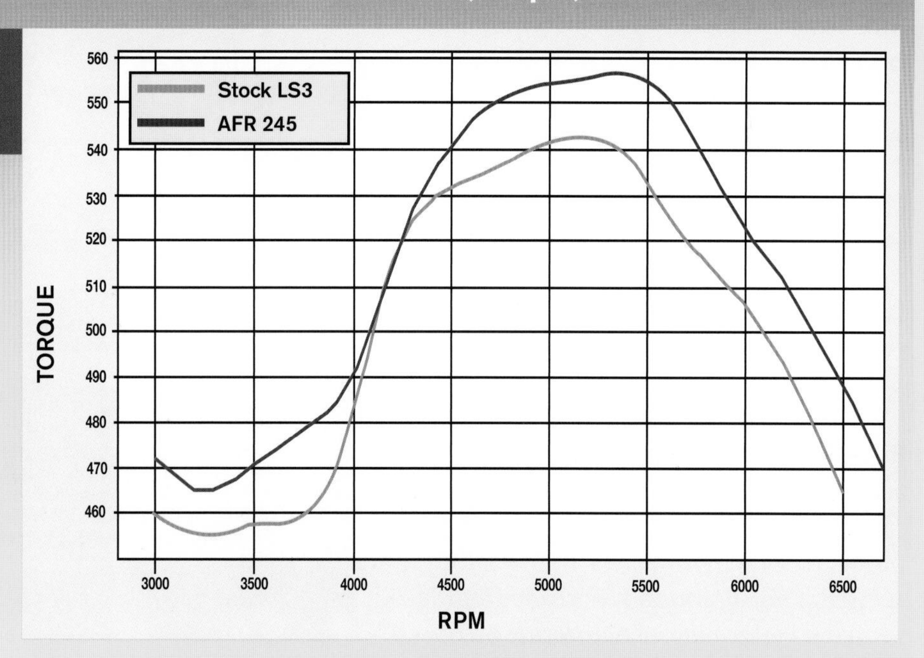

Test 3: Chevy Performance vs Brodix vs LPE on an LS7 495 Stroker

Even more so than with LS3 head testing, comparing LS7 heads (especially improved versions) requires a serious test engine. Even at 427 ci, the stock 7.0 isn't enough engine to tax the flow capacity of the stock LS7 heads, to say nothing of aftermarket heads that flow near 400 cfm. To properly test the merits of ported LS7 heads, I built a serious test mule that offered increased displacement, compression, and cam timing. Only then could I show what ported LS7 heads were really capable of.

CNC ported right from the factory, the stock LS7 heads were no slouch. Run on the 495 stroker, the stock heads produced more than 770 hp and more than 700 ft-lbs of torque.

Starting with an RHS tall-deck block, I bored out the engine to 4.185 inches then added a *big* stroker crank. Filling the stroker-friendly RHS block were equally stout components from Lunati, Wiseco, and K1. Lunati supplied a massive 4.5-inch billet stroker crank, which was combined with a set of 6.30-inch forged K1 rods and 4.185-inch Wiseco forged, flat-top pistons. The bore and stroke combined to produce a flow-taxing displacement of 495 ci.

The displacement and flat-top pistons combined with the 70-cc combustion chambers to produce a static compression ratio near 13.5:1. Sealing the beast was a set of Cometic MLS head gaskets secured by ARP heads studs. Additional components employed on the RHS stroker included a Moroso pan, pickup, and windage tray; a custom timing chain from Comp Cams designed specifically for the tall-deck block (1 extra link per side); and the largest off-the-shelf hydraulic roller cam available in the Comp Cams catalog. The 309LRR HR15 offered a .660 lift (with 1.8 rockers), 259/275-degree duration split, and 115-degree LSA.

Head swaps on the engine dyno were a snap, but it was necessary to prep the test by checking pushrod lengths for the various cylinder heads.

Feeding the beast was a Mast Motorsports (high-rise) single-plane intake designed to accept the Holley 1050 Ultra Dominator carb. Run with the stock LS7 heads, the 495 produced 773 hp and 704 ft-lbs of torque. Adding a set of ported LS7 heads from Lingenfelter Performance Engineering increased the power output to 793 hp and 719 ft-lbs of torque, while Brodix LS7 heads produced 799 hp and 714 ft-lbs of torque.

Chevy Performance vs Brodix vs LPE on an LS7 495 Stroker (Horsepower)

Chevy Performance CNC LS7 Heads: 773 hp @ 6,300 rpm
Brodix LS7 Heads: 793 hp @ 6,400 rpm
LPE CNC LS7 Heads: 799 hp @ 6,500 rpm
Largest Gain: 31 hp @ 6,600 rpm

Much like the LS3 head test run on the 468 stroker, this LS7 head test on the larger 495 stroker showed just how well the stock heads work. CNC ported right from the factory, the LS7 heads offered 773 hp on this RHS-block stroker. Installation of the ported heads from Brodix and LPE pushed the peak numbers near 800 hp, with gains as high as 31 hp.

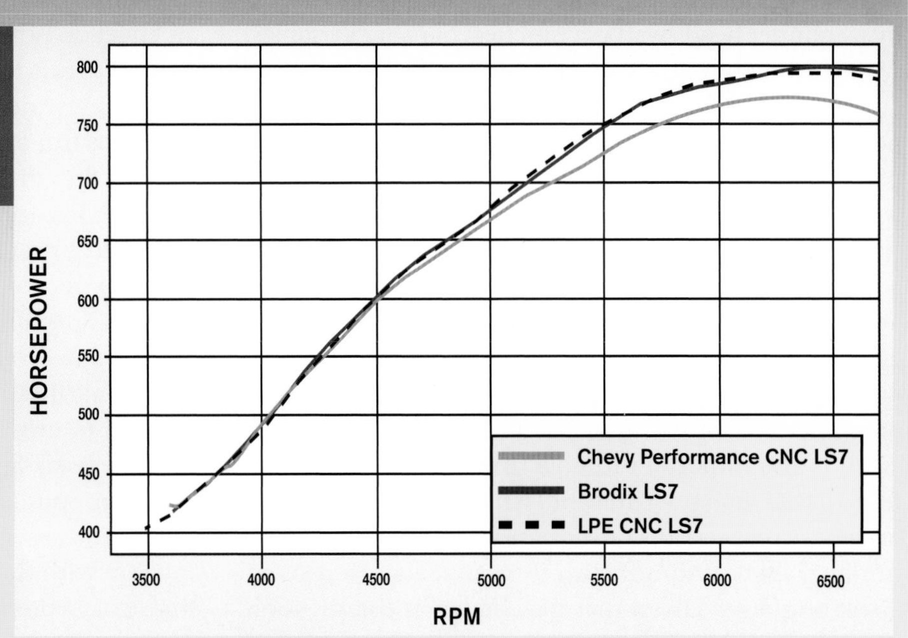

Chevy Performance vs Brodix vs LPE on an LS7 495 Stroker (Torque)

Chevy Performance CNC LS7 Heads: 704 ft-lbs @ 4,700 rpm
Brodix LS7 Heads: 719 ft-lbs @ 5,300 rpm
LPE CNC LS7 Heads: 714 ft-lbs @ 5,400 rpm
Largest Gain: 24 ft-lbs @ 5,400 rpm

The torque gains offered by the head swap on the 495 stroker came primarily past 4,500 rpm. There were minor gains before that point, but the extra airflow offered by the ported heads from Brodix and LPE made itself known higher in the rev range. The largest torque gain occurred at 5,400 rpm.

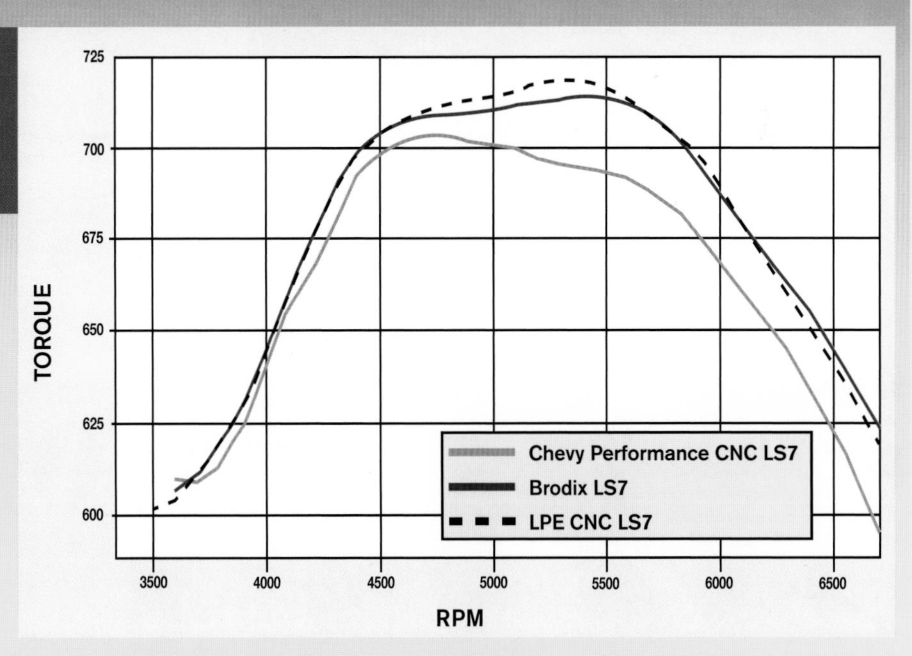

Test 4: Effect of Chamber Volume: TEA vs Speedmaster on an LS3

For the first three tests in this chapter, I have concentrated primarily on airflow, but the power offered by a cylinder head swap is a function of other variables. As mentioned in Test 2 on the AFR versus LS3 heads, a change in combustion chamber volume can alter the power output as well. To illustrate this, I selected two pairs of heads that offered similar flow rates but had a significant difference in chamber volume that substantially altered the static compression ratio.

Before I continue, it is important to know that an increase in static compression ratio (from 9.0:1 to 10.0:1, for instance) will increase the power output of any engine. The general rule is that the power increases 3 to 4 percent for each full point of compression.

It is also important to note that this rule is a guideline that is most accurate in what I call the normal range of compression ratios (8.0:1 to 13.0:1). Changes in ratios above and/or below this range have less of an effect on power. Given that the change in compression ratio came from increased chamber volume, the change in chamber shape (irrespective of size) might also have an effect on power production, but that requires a much more difficult test to prove.

This test was run on an LS3 crate engine upgraded to stroker status and equipped with a Texas Speed cam, long-tube headers, and two different cylinder heads.

This test on chamber volume (and static compression) was run on a 416 LS3 stroker. The stroker was built using a Speedmaster crank and rods combined with a set of JE flat-top pistons. Similar to many of the tests, I tried to lower the static compression to safely apply boost to the combination.

The Speedmaster crank, rods, and JE pistons were given a precision balance job and then assembled using Total Seal rings; Sealed Power bearings; and Fel Pro gaskets, oil pump, and timing chain. The stroker also featured a Texas Speed cam (.614/.621 lift split, 231/239 duration split, and 113 LSA) using Comp hydraulic roller lifters.

Run with the (big-chamber) Speedmaster LS3 heads, the stroker produced 576 hp and 551 ft-lbs of torque. After installation of the (smaller-chamber) Chevy Performance heads (with nearly identical flow numbers), the power output improved to 608 hp and 570 ft-lbs of torque. The change in compression ratio of 1.3 points improved the power output by 5.5 percent.

Both the Speedmaster and Chevy Performance heads featured full porting to enhance the flow rates.

Effect of Chamber Volume: TEA vs Speedmaster on an LS3 (Horsepower)

Speedmaster LS3 (big chamber) Heads: 576 hp @ 6,100 rpm
Chevy Performance LS3 (small chamber) Heads: 608 hp @ 6,200 rpm
Largest Gain: 35 hp @ 6,400 rpm

Changes in airflow typically improve power in relation to engine speed. The gains increase as the need for airflow increases. By comparison, a change in compression ratio improves power everywhere, from top to bottom. The small(er)-chamber Chevy Performance heads improved the power output of the LS3 stroker from 3,500 to more than 6,500 rpm.

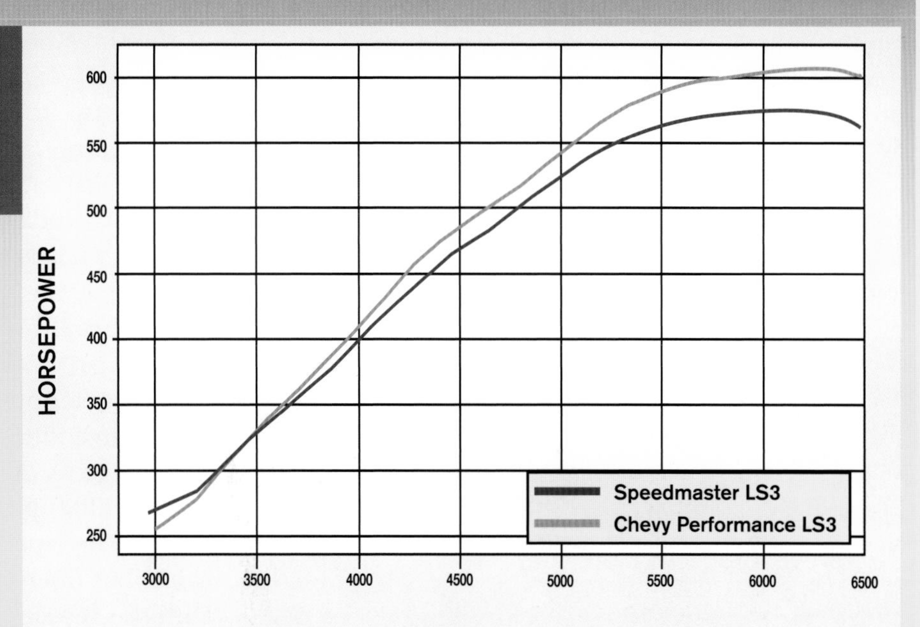

Effect of Chamber Volume: TEA vs Speedmaster on an LS3 (Torque)

Speedmaster LS3 (big chamber) Heads: 551 ft-lbs @ 5,100 rpm
Chevy Performance LS3 (small chamber) Heads: 570 ft-lbs @ 5,100 rpm
Largest Gain: 27 ft-lbs @ 5,600 rpm

Big torque gains are always welcome and that is exactly what increased compression provides. Of course, there is a limit to the gains offered by further increasing compression because the gains start to diminish past 13.0:1. The change in compression offered by the head swap netted more than 25 ft-lbs at 5,600 rpm.

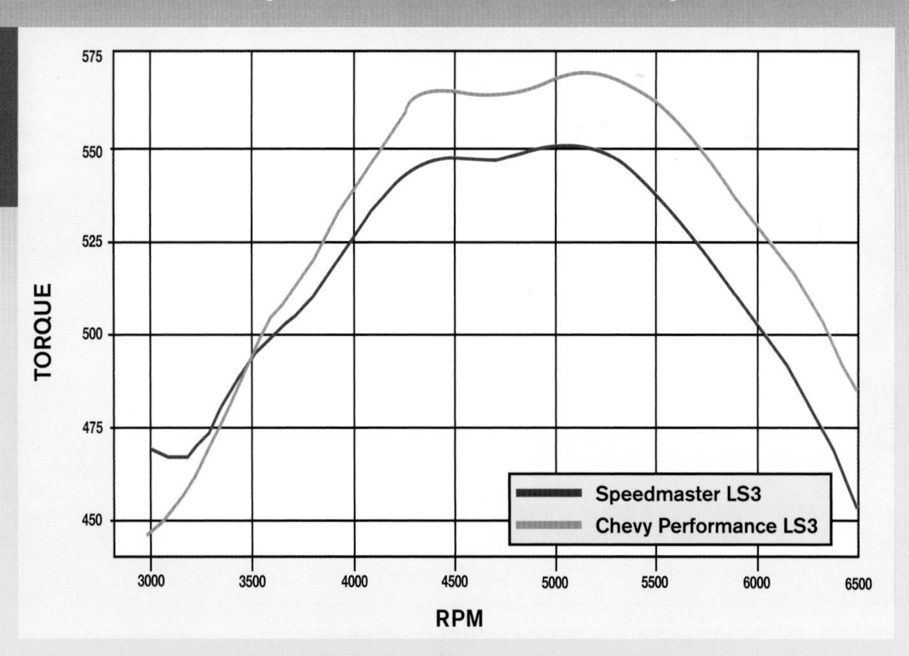

Test 5: Stock LS3 vs TEA vs Speedmaster LS3 on a 468 Stroker

I ran several ported LS3 heads on the 468 stroker engine. The problem is that I couldn't show all of the results in one test, so I elected to show this one.

As a recap, the 468 stroker used a Darton-sleeved LS6 block with forged internals from Lunati and JE. Aiding in power production was a healthy static compression ratio of 12.0:1. To allow the stroker to maximize power, I installed an off-the-shelf Comp cam (.624 lift, 255/271 duration split, 112 LSA) along with a set of Comp short-travel lifters. Milodon supplied a pan and remote oil filter to work with the modified windage tray (to clear the stroker). ARP and Fel Pro secured each of the three heads tested. Once again, the two ported LS3 heads were compared to a pair of ported versions, in this case from Total Engine Airflow (TEA) and GM Performance.

The LS6 aluminum block received Darton sleeves to allow me to bore out the block to 4.185 inches. The overbore was combined with a 4.25-inch Lunati stroker crank.

The first order of business was to run the 468 with the stock heads to establish a baseline. Equipped with stock LS3 heads, the stroker produced 692 hp at 6,500 rpm and 625 ft-lbs of torque at 4,900 rpm. Next up was a set of heads from TEA, which applied its Stage 2 porting to a set of factory castings. The porting improved the power output from 692 hp and 625 ft-lbs of torque to 719 hp and 631 ft-lbs.

The gains offered by the TEA were most prevalent at high RPM, but there were gains through the entire range. The Speedmaster LS3 heads showed similar gains, but offered slightly more peak power. Equipped with the Speedmaster LS3 heads, the 468 produced 730 hp, but peak torque dropped slightly to 629 ft-lbs.

This testing tells me that the stock LS3 heads are plenty powerful and that a number of different aftermarket heads can offer substantial power gains.

The Total Engine Airflow (TEA) LS3 heads featured full porting and a valvespring upgrade. The Stage 2 porting increased the flow rate from 314 to 365 cfm.

Stock LS3 vs TEA vs Speedmaster LS3 on a 468 Stroker (Horsepower)

Stock LS3 Heads: 692 hp @ 6,500 rpm
TEA LS3 Heads: 719 hp @ 6,900 rpm
Speedmaster LS3 Heads: 730 hp @ 6,900 rpm
Largest Gain: 46 hp @ 7,000 rpm

Compared to the stock LS3 heads, the gains offered by the TEA and Speedmaster heads increased with engine speed. The gains were greatest at 7,000 rpm, especially with the heads from Speedmaster. Where the stock heads started falling off in power, the ported heads were just getting started.

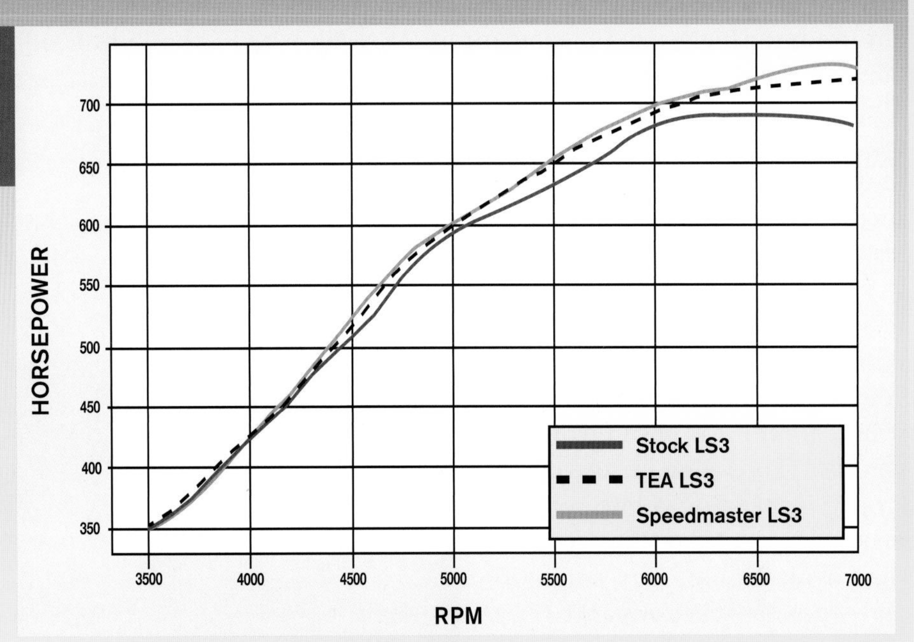

Stock LS3 vs TEA vs Speedmaster LS3 on a 468 Stroker (Torque)

Stock LS3 Heads: 625 ft-lbs @ 4,900 rpm
TEA LS3 Heads: 631 ft-lbs @ 4,900 rpm
Speedmaster LS3 Heads: 629 ft-lbs @ 4,900 rpm
Largest Gain: 23 ft-lbs 5,600 rpm

The torque gains didn't really materialize at engine speeds less than 4,400 rpm. This is where the flow rate of the stock heads was more than sufficient to feed the power needs of the 468 stroker. The torque gains increased with engine speed, but the gains were abundant from 4,500 to 7,000 rpm.

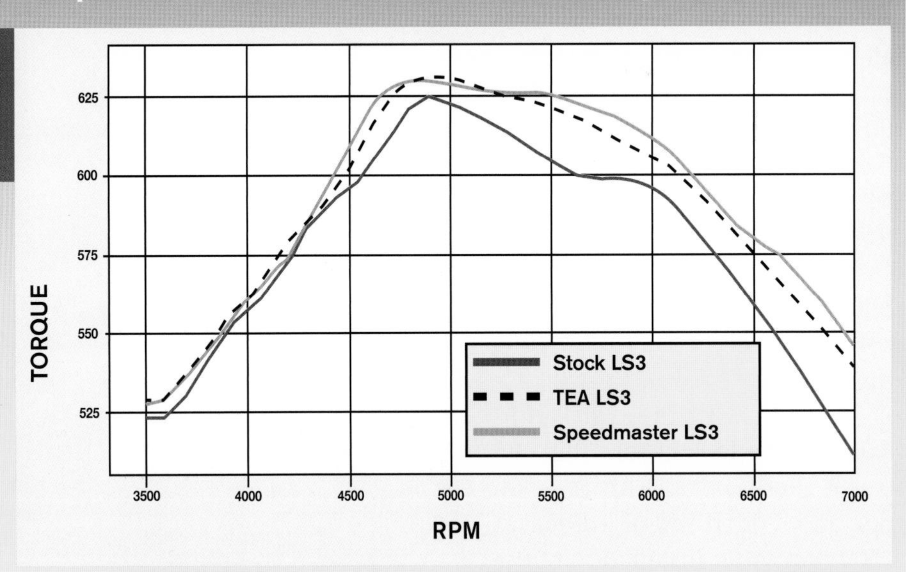

Test 6: Chevy Performance vs TS vs SDPC on an LS7 495 Stroker

The testing performed on the LS7 heads paralleled the testing on the LS3 heads in that I ran several different heads during the dyno comparison. After all, why take the time to set everything up multiple times when you have the test engine ready and all you have to do is swap the heads? In truth, there is much more to this type of testing than simple head swaps.

Every head must be first mocked up to measure for proper pushrod length. The last thing you want to do during testing is search around for pushrods when you have an engine on the dyno. Proper preparation allows you to maximize available (and expensive) dyno timing. This test compares the Chevy Performance CNC LS7 heads (basically stock) and two sets from Texas Speed (TS) and Scoggin Dickey Performance Center (SDPC).

Once again, I used the 495-inch stroker with an RHS tall-deck block, Lunati billet crank K1 forged rods, and Wiseco pistons. The 309LRR HR15 featured a .660-inch lift (with 1.8 rockers), 259/275-degree duration split, and 115-degree LSA. The induction system included a Mast Motorsports (high-rise) single-plane intake designed to accept the Holley 1050 Ultra Dominator carb.

First run with the GM Performance CNC LS7 heads, the 495 stroker produced 773 hp at 6,300 rpm and 704 ft-lbs of torque at 4,700 rpm. After installation of the Texas Speed LS7 heads, the power output jumped to 796 hp and 725 ft-lbs of torque. Credit a peak flow rating of 403 cfm offered by the TS heads for the big power gains. The SDPC LS7 heads offered 389 cfm, which allowed them to produce 802 hp, but torque fell slightly (compared to the TS heads) to 717 ft-lbs.

When you are trying to make more displacement, nothing beats a tall-deck, RHS aluminum block. The tall-deck RHS block allowed me to bore and stroke the LS to 495 ci.

The RHS block featured a dedicated cam retaining plate. The 495 featured a healthy (off-the-shelf) Comp hydraulic roller cam.

Chevy Performance vs TS vs SDPC on an LS7 495 Stroker (Horsepower)

Chevy Performance CNC LS7 Heads: 773 hp @ 6,300 rpm
Texas Speed CNC LS7 Heads: 796 hp @ 6,200 rpm
SDPC CNC LS7 Heads: 802 hp @ 6,500 rpm
Largest Gain: 39 hp @ 6,700 rpm

As I have come to expect from head porting, the power gains increased with engine speed. Equipped with factory (Chevy Performance) LS7 heads, the 495 produced 773 hp. The TS heads pushed the peak power to 796 hp; the SDPC heads produced 802 hp. The TS heads offered slightly more power through the rev range, but fell off on top compared to the SDPC heads.

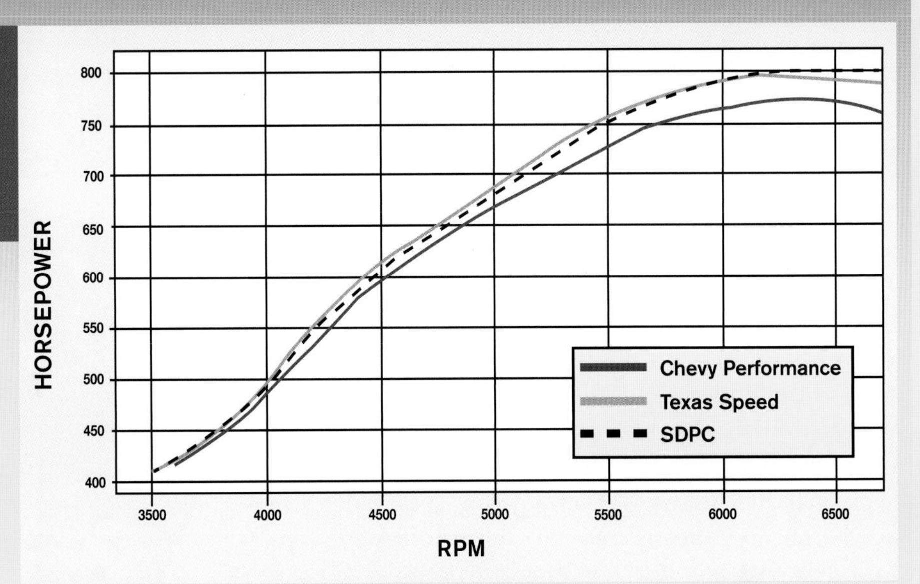

Chevy Performance vs TS vs SDPC on an LS7 495 Stroker (Torque)

Chevy Performance CNC LS7 Heads: 704 ft-lbs @ 4,700 rpm
Texas Speed CNC LS7 Heads: 725 ft-lbs @ 5,200 rpm
SDPC CNC LS7 Heads: 717 ft-lbs @ 5,400 rpm
Largest Gain: 30 ft-lbs 5,400 rpm

The ported heads offered impressive torque gains, with the greatest gain coming at 5,400 rpm. The bump in torque production offered by the TS heads coincided with the dip in torque with the Chevy Performance LS7 heads.

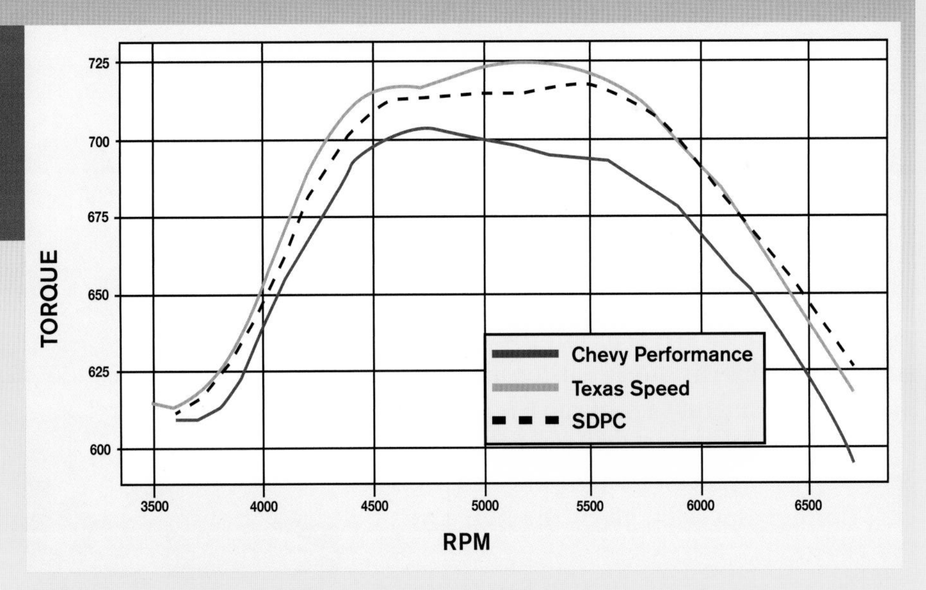

Test 7: Stock LS3 vs TFS Gen X 255 on a Modified LS3

One of the critical components of any LS combination is the cylinder heads. As luck (or design) would have it, the LS3 was blessed with plenty of head flow right from the factory, but that doesn't mean the stock heads can't be improved on. Case in point, the Trick Flow Gen X 255 LS3 heads. According to Trick Flow, the Gen X heads combined a peak airflow of 380 cfm with an intake port volume that measured just 255 cc. That is where the Gen X 255 heads get their name. The port volume is important because it's easy to make big flow with big ports. The key to a successful cylinder head is to maximize flow while minimizing port volume. To put the TFS numbers into perspective, the stock LS3 heads checked in with just 315 cfm from 260-cc ports. The TFS heads offered significantly more flow *and* less port volume, a true indication of a solid design. The Trick Flow heads also offered a spring package that allowed me to run a healthy camshaft; after all, why upgrade heads on an otherwise stock engine? The TFS heads also featured a spring package that offered 150 pounds of seat pressure and 400 pounds of open pressure.

For a test engine, I once again relied on the LS3 crate engine from Gandrud Chevrolet. Prior to the test, the LS3 was upgraded with a hot Crane hydraulic roller cam. The Crane cam offered a .600-inch lift (intake and exhaust), 232/240-degree duration split (at .050), and 113-degree LSA. This cam was nearing the limit of available piston-to-valve clearance offered by the stock flat-top pistons, or more specifically the lack of valve reliefs. With proper valve reliefs, additional power would certainly be available with more aggressive cam timing, especially given the flow rate of the TFS heads.

The LS3 was run with a Holley HP management system and FAST injectors. With stock heads, the modified LS3 combination produced 552 hp at 6,400 rpm and 513 ft-lbs of torque at 5,000 rpm. After installation of the TFS Gen X 255 heads, the power numbers jumped to 571 hp at 6,500 rpm and 525 ft-lbs of torque at 5,200 rpm. As with other tests conducted in this chapter, the greater the power output of the test engine, the greater the gains with a head swap.

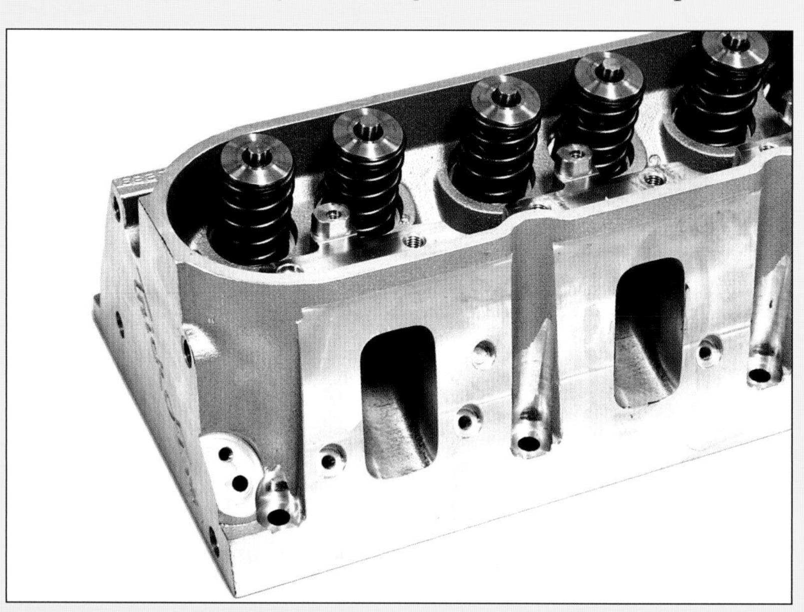

The TFS Gen X 255 heads featured amazing flow (more than 380 cfm)* and *smaller-than-stock port volumes. Also present was a spring package that allowed me to run the heads with ample camshaft to make power.

Bolting the right set of heads on your modified LS3 is a surefire route to improved performance.

Stock LS3 vs TFS Gen X 255 on a Modified LS3 (Horsepower)

Stock LS3 Heads: 552 hp @ 6,400 rpm
TFS Gen X 255 Heads: 571 hp @ 6,500 rpm
Largest Gain: 21 hp @ 6,500 rpm

The head swap was worth plenty of power, but this mild combination could not take full advantage of what the TFS heads had to offer. The flow rate suggests the head can feed an NA combination near 800 hp, but on the 550-hp LS3, the head swap was worth near 20 extra horsepower.

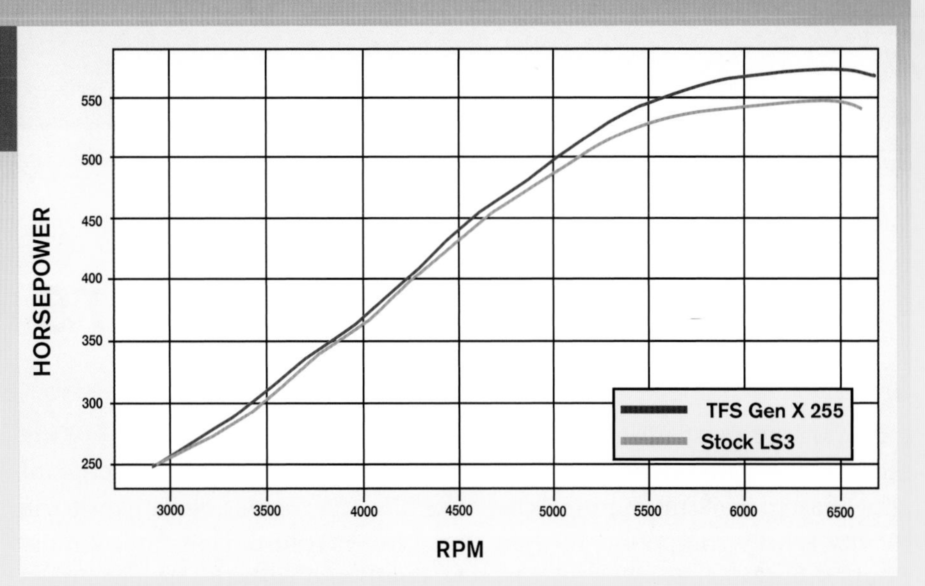

Stock LS3 vs TFS Gen X 255 on a Modified LS3 (Torque)

Stock LS3 Heads: 513 ft-lbs @ 5,000 rpm
TFS Gen X 255 Heads: 525 ft-lbs @ 5,200 rpm
Largest Gain: 17 ft-lbs 5,500 rpm

I liked the fact that the head swap improved torque production through the entire rev range, even if only by a little down low. Gains on the positive side are always better than having to trade low-speed torque for high-RPM power. Because the stock LS3 heads easily support the power level of this mild combo, the gains offered by the head swap were not what they could be. What these heads need is a 468 or 495.

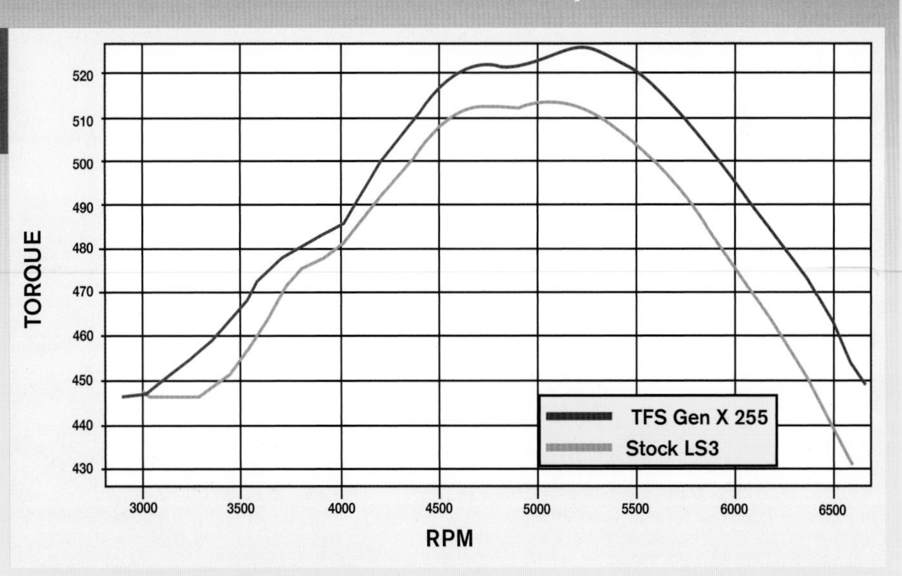

CAMSHAFTS

One of the most misunderstood performance components on any engine has to be cam timing. The difficulty is only compounded when you add things such as nitrous oxide, turbos, or superchargers. From an anatomical standpoint, the camshaft can be likened to the brain because the cam profile determines how effectively (when and where) breathing takes place.

The camshaft is one of the major determining components of the effective operating range of the engine. Of course, cam timing must be combined with the proper intake manifold, head flow, and primary length on the exhaust for optimum operation over a given RPM range, but the right cam can almost determine the character or personality of the engine. Stock or ultra-mild aftermarket cams provide a dead-smooth idle, while more radical grinds can transform that mild-mannered LS engine into one radical ride.

The single best (and most powerful) modification you can make to your LS3 or LS7 engine is a cam swap. These engines already feature a good intake and plenty of head flow, so all they need to make amazing power is more aggressive cam timing.

The factory LS3 is a common upgrade for 4.8 and 5.3 applications because those lesser LS engines were equipped with the mildest factory cams ever offered. Of course, the high-performance LS7 was factory equipped with the most powerful cam ever offered on an LS, but the stock stuff is just begging to be replaced.

The LS3 and LS7 are fantastic engines, offering an impressive combination of power, reliability, and even fuel mileage. Another area where they excel is how well they respond to performance upgrades, especially camshafts. These factory performance engines respond so well to wilder cam timing because they have everything else required to make power, including displacement, intake, and head flow. All that is lacking to dramatically improve the power output of a typical LS3 or LS7 is cam timing.

With the factory heads and intake already capable of supporting more than 600 hp (700 hp on the LS7), mild cams are definitely the limiting factor. Given this situation, cam upgrades for LS applications have become hot sellers. Plop just about any cam in an otherwise stock LS and watch the power soar. I have seen power gains of 65 to 70 hp from a simple cam swap on an otherwise stock LS application. The gains can be even greater higher in the rev range.

There is, of course, a limit to how wild you can go with cam timing on an otherwise stock LS (3 or 7) engine. Although the LS duo certainly responds to more aggressive cam timing, two limitations are inherent in the stock combinations. First, stock valvesprings were designed for stock cams and are, therefore, insufficient for performance use. From available valve lift and RPM potential standpoints, spring swaps are not just a good idea, they should be considered mandatory for most cam upgrades on an LS3 or LS7.

The other limiting factor in terms of cam timing on a stock LS3 or LS7 application is available piston-to-valve clearance. Although lift plays a minor role, the real culprit in piston-to-valve clearance is duration (how long you hang that valve open). Each successive increase in duration (the intake hits before the exhaust) decreases the available clearance. Cams that exceed 230 degrees of intake duration should always be checked, especially if they were ground with a few extra degrees of advance.

One of the most common questions regarding camshafts is which one is right for your LS combination. The term "right" here obviously has different meanings for different people, so choosing the so-called right cam can be difficult for even cam experts. The difficulty comes not in the technical nature of cam timing or profiles, but in deciphering exactly what you want. This becomes even more difficult when you are unsure.

Asking enthusiasts what cam they want is a little like asking them how much power they want. The problem is that they want as much power as possible and they also want a factory-smooth idle, 50 mpg, and maintenance-free operation. It goes without saying that it is not possible to combine all of those elements. Obviously that is a lot to ask of any camshaft and, ultimately, trade-offs become necessary. The question then becomes how many of the trade-offs you are willing to accept in your quest for power. Luckily for LS3 and LS7 owners, it doesn't require much in the way of cam timing to make a major difference in power.

LS cams are available in single- and three-bolt configurations. If you are upgrading an LS3 and select a three-bolt cam, make sure you have the matching three-bolt cam gear.

Factory hydraulic roller lifters work well, but there are aftermarket performance units available, including retrofit, short travel, and even solid roller versions.

The factory LS3 and LS7 feature offset intake rockers. The LS3 shares the 1.7-ratio, exhaust rocker with cathedral-port LS applications, but the 1.8-ratio LS7 is specific to that cylinder head configuration.

Test 1: Stock LS3 vs Comp Cams 281LRR on a Modified LS3

Cam swaps are popular for the LS family, especially the LS3, for good reason. Nothing adds power to an LS3 like a cam swap. Short of power adders or a stroker engine, no modification equals the power gains offered by a cam swap. This is because an LS3 already has sufficient displacement, intake, and (especially) cylinder head flow to make serious power. The only thing missing from the combination is cam timing. Add the right cam to an LS3 and watch the power needle climb.

The power gains are even more impressive when you further increase the power potential with ported LS3 heads such as the ones from Chevy Performance run on this LS3 crate engine. Chapter 2 illustrated that an added ported head to a stock engine offers very little in the way of extra power, but adding a cam to a combination with ported heads shows big gains.

This test engine was a GM Performance LS3 crate engine from Gandrud Chevrolet upgraded with GM Performance CNC L92 heads, ARP head studs, and a manual throttle body. The LS3 also featured Lucas 5W-30 synthetic oil, long-tube headers, and a Holley Dominator EFI management system.

Run first with the stock LS3 cam, the LS3 produced 503 hp at 5,500 rpm and 497 ft-lbs of torque at 4,600 rpm. The GM Performance L92 heads were supplied with stock LS3 springs, so it was necessary to install a set of dual springs from BTR to test the cam. After replacing the stock LS3 cam (and springs) with the Comp 281LRR cam (.617/.624 lift split, 231/239 duration split, and 114 LSA), the peak numbers jumped to 569 hp at 6,500 rpm and 522 ft-lbs of torque at 5,200 rpm. There was little change in power below 4,000 rpm, but the gains increased thereafter with engine speed.

The stock LS3 cam featured a single-bolt for the cam gear, but the Comp cam was a three-bolt design (Comp offers single-bolt cams as well). The three-bolt cam requires a 4X, three-bolt cam gear.

Here are the stock, single-bolt, 4X LS3 cam gear; the 4X three-bolt (LS2) gear; and an early 0X truck gear (cam sensor in rear of cam).

Stock LS3 vs Comp Cams 281LRR on a Modified LS3 (Horsepower)

Stock LS3 Cam: 503 hp @ 5,500 rpm
Comp 281LRR Cam: 569 hp @ 6,500 rpm
Largest Gain: 70 hp @ 6,400 rpm

The most amazing thing about the power generated by the Comp cam swap was not that it added a ton of power (it did), but that the amazing gains came with no loss in power at 3,000 rpm. The gains from the cam swap increased with engine speed and peak with 70 hp at 6,400 rpm.

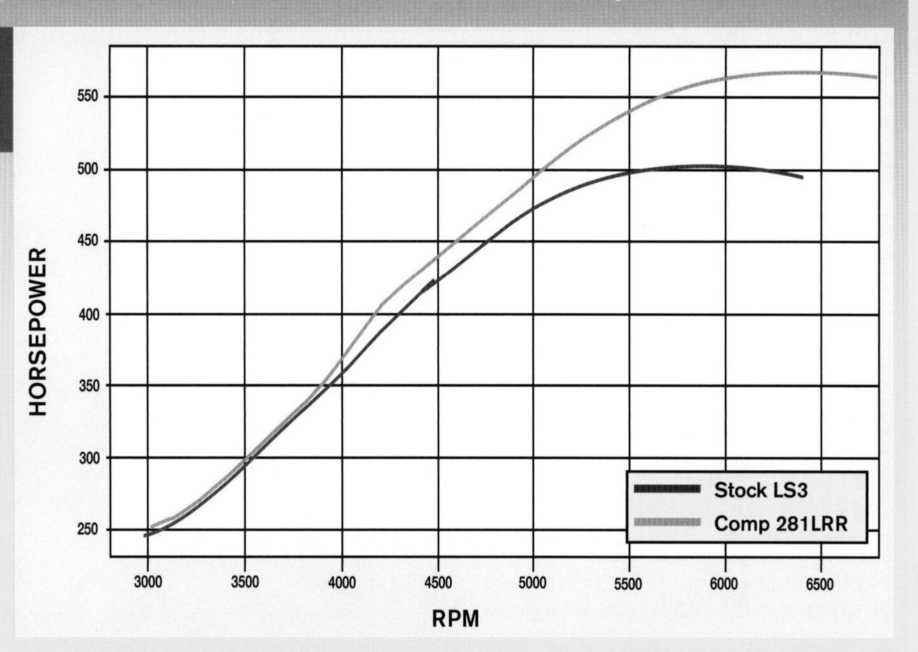

Stock LS3 vs Comp Cams 281LRR on a Modified LS3 (Torque)

Stock LS3 Cam: 497 ft-lbs @ 4,600 rpm
Comp 281LRR Cam: 522 ft-lbs @ 5,200 rpm
Largest Gain: 57 ft-lbs @ 6,100 rpm

What I like about adding a cam to an LS3 combination is that the amazing top-end power gains come with no penalty in low-speed torque. In fact, torque gains occurred just below 4,000 rpm and increased with engine speed. Larger cam profiles start to trade off low-speed torque for possible gains in peak power.

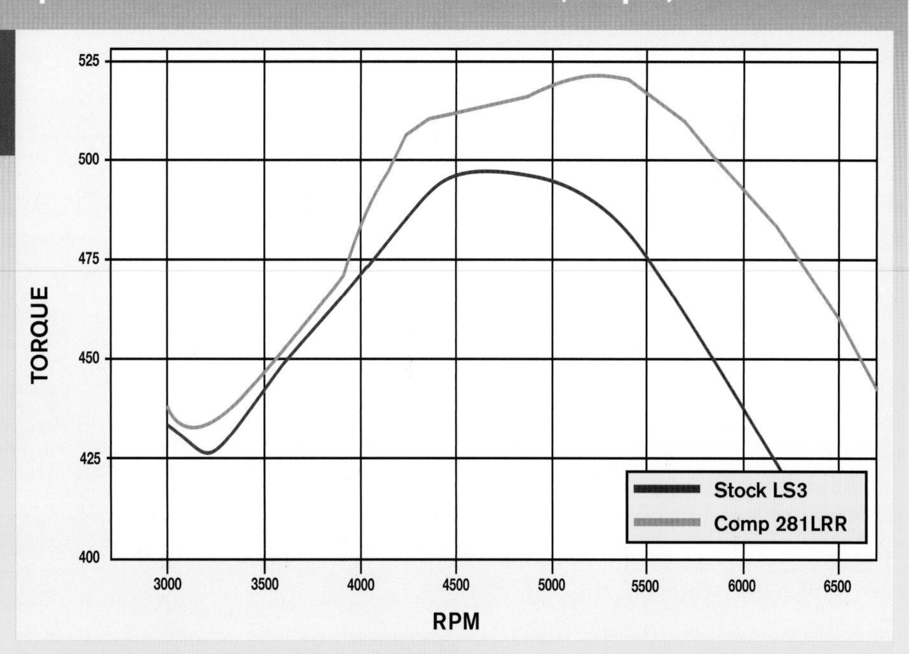

Test 2: Stock LS3 vs BTR Stage IV on an LS3

Adding just about any performance cam (and springs) to a stock LS3 is going to get you a lot of extra power, but adding the right cam can offer even more. To illustrate the gains possible with a cam-only upgrade on a stock LS3, I installed the Gandrud Chevy LS3 crate engine on the dyno and treated it to a Stage IV cam from BTR. Right at the limit of available piston-to-valve clearance, the Stage IV cam was the perfect candidate to work with the stock LS3 heads.

Remember, even in stock trim, an LS3 has an excess of cylinder head flow. The stock heads are capable of supporting nearly 700 hp on the right application (see Chapter 2), so the only thing missing in the combination is cam timing. It is also important to remember that not all cam-only upgrades are created equal and that most (like this one) must be combined with appropriate valvesprings (these came from BTR as well).

The Gandrud crate LS3 was installed on the engine dyno and run with the stock cam using long-tube headers, a Holley HP management system, and Lucas oil. Also present was a FAST (manual) throttle body, Meziere electric water pump, and K&N oil filter. Run with the stock LS3 cam, the LS3 produced 496 hp at 5,800 rpm and 488 ft-lbs of torque at 4,700 rpm.

Replacing the stock cam with the Stage IV from BTR also required swapping out the single-bolt, 4X cam sprocket for a three-hole, 4X version. After installation of the new cam, the power output jumped to 570 hp at 6,500 rpm and 522 ft-lbs of torque at 5,300 rpm. Not only was peak power production way up, but the cam swap netted torque gains all the way down to 3,000 rpm. Below 4,500 rpm, the BTR cam offered an extra (and consistent) 20 to 25 ft-lbs of torque, but this number increased substantially above 4,500 rpm.

The stock LS3 (flat-top) piston features no intake or exhaust valve reliefs. Available piston-to-valve clearance is the limiting factor in terms of cam timing on a stock LS3.

Although BTR offers cam profiles that require no spring upgrades, this Stage VI cam was combined with a dual-spring upgrade.

Stock LS3 vs BTR Stage IV on an LS3 (Horsepower)

Stock LS3 Cam: 496 hp @ 5,800 rpm
BTR Stage IV Cam: 570 hp @ 6,500 rpm
Largest Gain: 90 hp @ 6,600 rpm

As much as I liked the huge peak power gains (74 hp) and extra 90 hp at 6,600 rpm, I also liked the fact that the BTR Stage IV cam offered gains down low. You know you have made the right cam choice for your application when you get huge power gains with no trade-offs in low-speed torque.

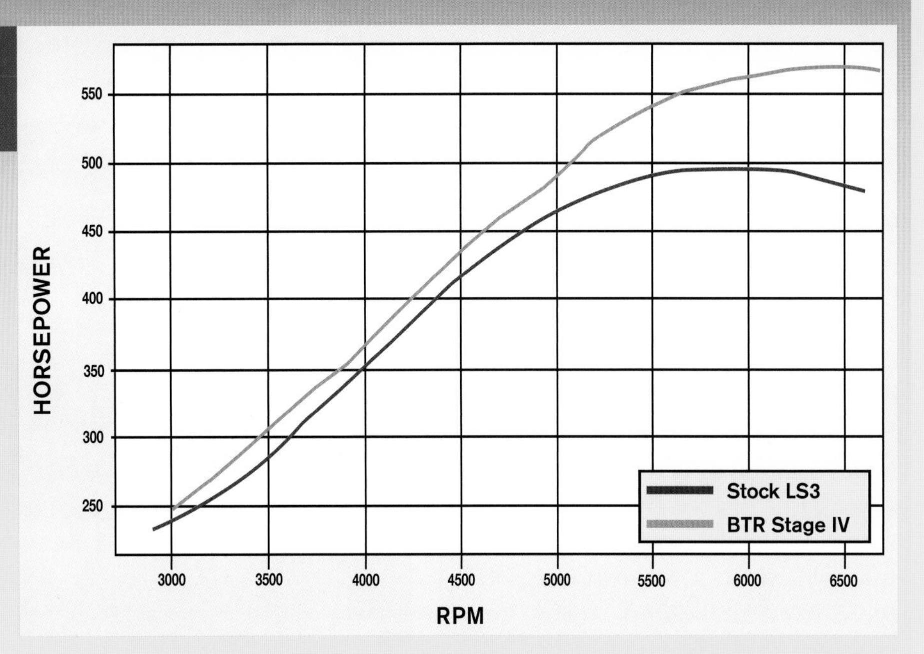

Stock LS3 vs BTR Stage IV on an LS3 (Torque)

Stock LS3 Cam: 491 ft-lbs @ 4,700 rpm
BTR Stage IV Cam: 522 ft-lbs @ 5,300 rpm
Largest Gain: 67 ft-lbs @ 6,400 rpm

Adding as much as 90 hp to an LS3 is an amazing thing, but it is the extra 25 ft-lbs down at 3,700 rpm that will be used more often in daily street driving. The extra 20–25 ft-lbs of torque up to 4,500 rpm will be most helpful in getting this LS3 up on the cam.

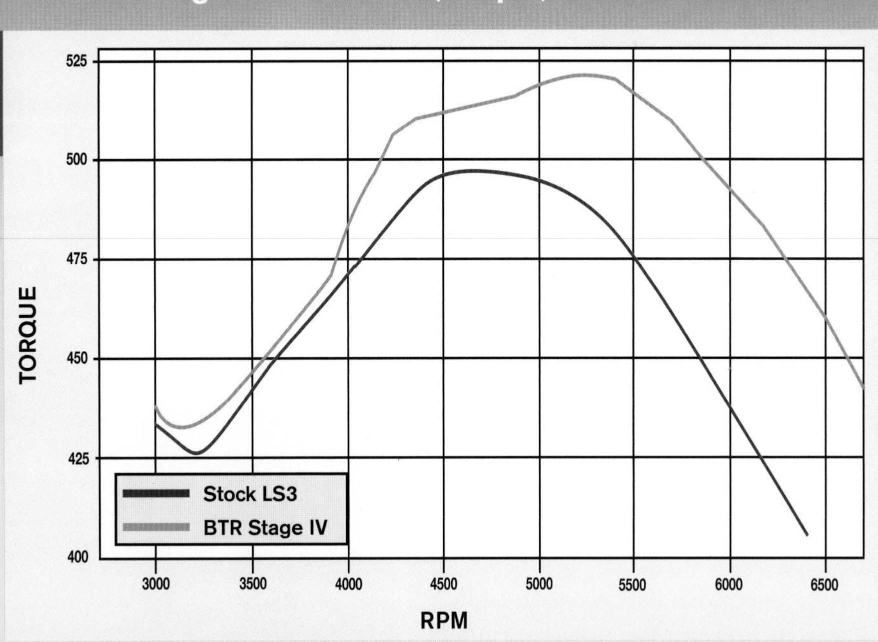

Test 3: LS9 vs LJMS Stage 2 Turbo on a Short-Stroke LS3

As you saw in the first two tests, cam timing is important for LS3 applications, but nowhere is it more important than on a turbo application. Spool-up of the turbo is a function of the power output of the engine at the desired spool RPM. The greater the power output, the quicker the spool-up. Things that can alter spool-up include displacement, cam timing, and intake design.

Small-displacement engines have more difficulty spooling up turbos (especially large ones), as do wilder cam timings (that may sacrifice low-speed torque) and short-runner intakes (that also reduce torque production lower in the rev range). This test illustrated the power gains offered by a cam swap on a short-stroke, LS3 turbo combination equipped with a short-runner, Holley Hi-Ram intake and large, 76-mm Precision turbo (meaning it had three strikes against it in terms of spooling). Thus, cam selection was even more critical on this engine than a stock LS3.

The short-stroke turbo engine featured an aluminum LS3 block equipped with a 4.8 crank, 6.30-inch forged Lunati rods, and custom JE pistons. The combination also included Total Seal rings, TFS Gen X 255 heads, and a Holley Hi-Ram intake (with 102-mm FAST throttle body). Rounding out the package was a Moroso oiling system, ATI dampener, and custom DNA turbo manifolds. The manifolds fed a single Precision 76-mm turbo, a CX Racing ATW intercooler, and TurboSmart 45-mm Hypergate waste gates.

The engine was first run (on the waste-gate springs) with an LS9 cam. So equipped, the turbo combination produced 697 hp and 598 ft-lbs of torque. After installation of the LJMS Stage 2 Turbo cam, the power numbers jumped to 733 hp and 621 ft-lbs of torque, but the real story is how much extra low-speed torque the cam swap offered. The gains would be even greater had I elected to equalize the boost pressure because the boost dropped by .5 psi after the cam swap (I ran it on the spring).

The test engine was an LS3 block equipped with a 4.8 crank, 6.3-inch Lunati forged rods, and JE pistons. The short-stroke LS3 was topped with a set of TFS Gen X 255s, a Holley Hi-Ram intake, and a custom turbo kit that featured stainless manifolds from DNA feeding a Precision 76-mm turbo.

Adding a turbo to the max-performance build makes cam selection even more important.

LS9 vs LJMS Stage 2 Turbo on a Short-Stroke LS3 (Horsepower)

LS9 Cam: 697 hp @ 6,900 rpm
LJMS Stage 2 Turbo Cam: 733 hp @ 6,600 rpm
Largest Gain: 34 hp @ 6,600 rpm

In some instances, a cam swap offers substantial power gains at higher engine speeds. Replacing the most powerful factory cam available (the LS9) with an LJMS Stage 2 turbo cam resulted in peak power gains as well as gains through the entire rev range. The cam swap also dropped the boost pressure by as much as .5 psi. More power with less boost is always a good thing.

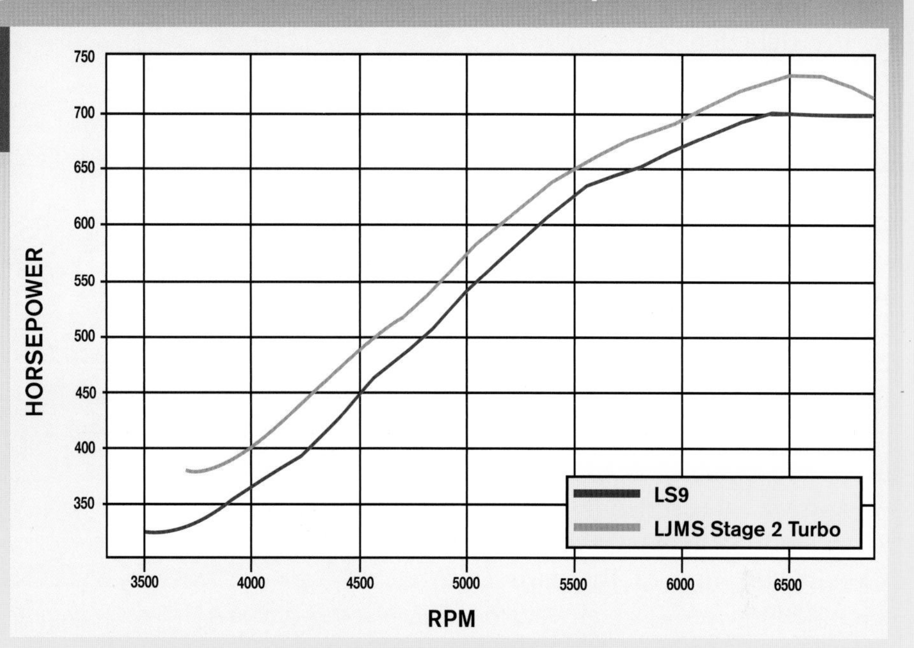

LS9 vs LJMS Stage 2 Turbo on a Short-Stroke LS3 (Torque)

LS9 Cam: 598 ft-lbs @ 5,600 rpm
LJMS Stage 2 Turbo Cam: 621 ft-lbs @ 5,500 rpm
Largest Gain: 61 ft-lbs @ 3,800 rpm

The extra 34 hp was obviously welcome, but it was the extra 60 ft-lbs down low that really made this LJMS cam a success. By increasing the low-speed torque production (actually through the entire curve), the LJMS cam would certainly offer increased boost response on the street (or strip). The artificial load from the engine dyno negates some of the gain, but boost response with the turbo cam would be greatly enhanced with an extra 60 ft-lbs on tap.

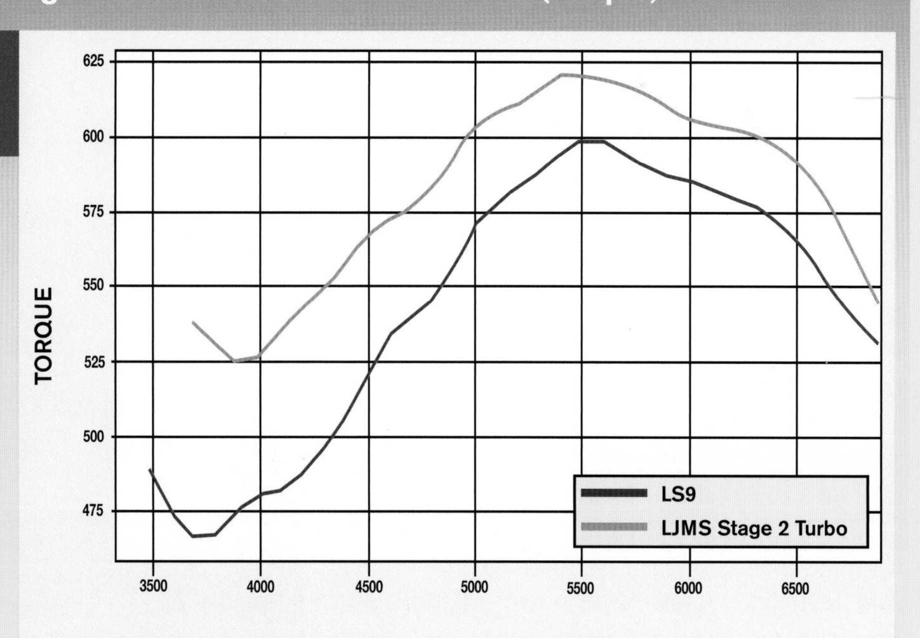

Test 4: NA vs BTR Stage IV Blower Cam on an SC LSX

Building a performance LS engine means using the right tool for the right job. This applies to more than just a torque wrench or feeler gauges; it applies to running the proper cam grind for the intended use and application. In this case, the application was a supercharged B15 LSX engine. Built by General Motors to withstand 15 psi (the 15 in B15), the forced-induction combo (not just blowers) featured a forged crank and pistons (powdered metal rods) along with six-bolt LSX LS3 heads.

Add a boost-friendly static compression ratio of 9.0:1 and you have the makings of the perfect combination for a supercharged cam test. All I did was add a Whipple supercharger, 150-pound injectors, and a Holley HP management system and I was ready to roll.

This test was run on a B15 crate engine equipped with a 3.3 Whipple supercharger.

This test was designed to compare an NA cam to a dedicated blower grind on a supercharged LS3 application. In addition to being designed for a positive-displacement blower application, the Stage IV cam from BTR was also slightly more aggressive. The NA cam (from Comp Cams) spec'd out with a .617/.624-inch lift split, 231/247-degree duration split, and 113-degree LSA. By comparison, the BTR blower cam offered the same lift split, an extra 8 degrees of intake duration, 11 degrees of exhaust duration, and 6 degrees of LSA (113 vs 119).

Equipped with the NA cam, the Whipple-supercharged LSX produced 855 hp at 6,700 rpm and 713 ft-lbs of torque. After installation of the BTR Stage IV cam, the peak numbers jumped to 880 hp and 714 ft-lbs. The milder NA cam actually offered more power up to 4,800 rpm, but the BTR cam pulled away thereafter. Interestingly, the boost was higher with the BTR cam than with the NA cam, despite no change in pulley size.

Even with the blower, the cam swap was easy. Run with the (plenty powerful) NA cam, the supercharged combo produced 855 hp and 713 ft-lbs of torque.

NA vs BTR Stage IV Blower Cam on an SC LSX (Horsepower)

NA Cam: 855 hp @ 6,700 rpm
BTR Blower Cam: 880 hp @ 6,700 rpm
Largest Gain: 29 hp @ 6,500 rpm

Although designed for an NA (rectangular-port) application, the Comp cam offered plenty of power on the supercharged LSX. The tighter LSA and shorter duration on the NA cam offered more power down low on the supercharged applications, but the wilder Stage IV BTR cam came on strong at the top of the rev range.

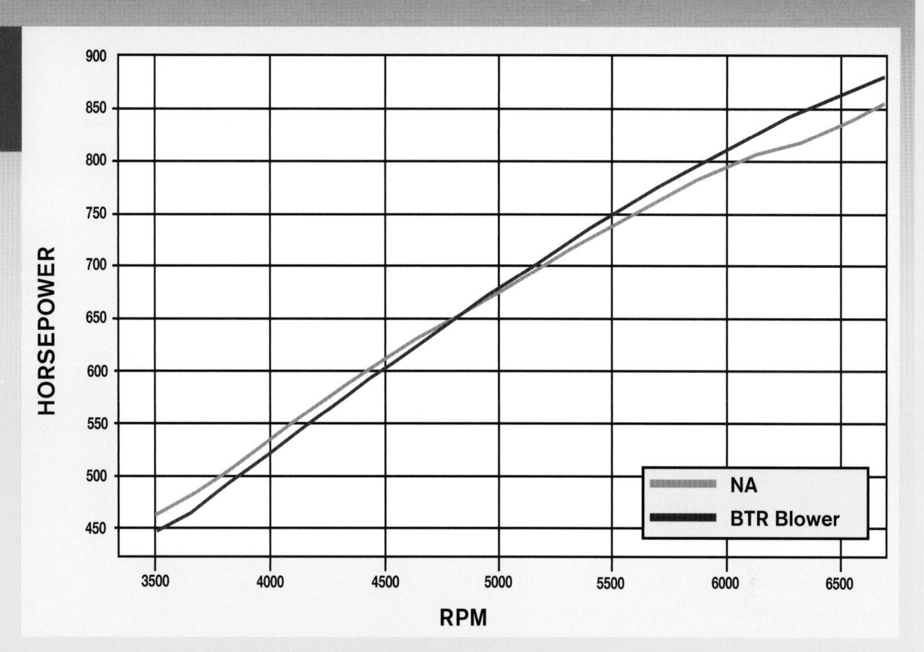

NA vs BTR Stage IV Blower Cam on an SC LSX (Torque)

NA Cam: 713 ft-lbs @ 4,600 rpm
BTR Blower Cam: 714 ft-lbs @ 5,300 rpm
Largest Gain: 23 ft-lbs @ 3,600 rpm

The torque curve shows what might be considered typical of a sizable change in cam duration. Stepping up 8 degrees in cam duration and increasing the LSA by 6 degrees (from 113 to 119) enhanced top-end power production. There was a trade-off in torque production below 4,700 rpm because the milder (NA) cam offered better power down low.

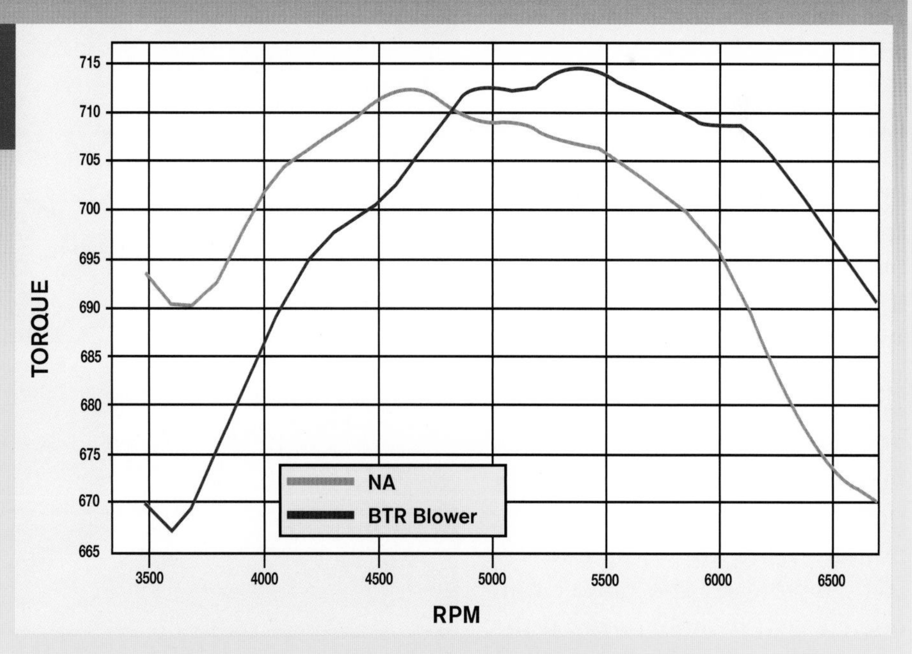

Test 5: Effect of LSA on a Supercharged LSX

Does the LSA affect the boost and power curves of a supercharged LS3? Obviously the answer is yes or this would be one very short test. Typically blower cams (such as the factory LSA and LS9 cams) are ground with very wide LSAs. Those factory cams featured 121.5-degree LSAs, but it is not uncommon for blower cams to be in the 119- to 120-degree area. In comparison, typical performance cams for LS3 applications are slightly tighter, in the 112- to 114-degree range.

This begs two questions: What happens if you just run your blower engine with a cam that you know works well on an NA LS3? Will a powerful LS3 cam work well once you add a supercharger?

To answer these questions, I set up a test on the GM B15 LSX engine. Equipped with a Whipple supercharger, I wanted to test two cams with distinctly different LSAs. The two cams (one from Comp and the other from BTR) were as close as possible in specs other than the LSA. The tight LSA cam from Comp had a .617/.624-inch lift split, 231/247-degree duration split, and 113-degree LSA. The BTR blower cam had a .617/.596-inch lift split, 231/248-degree duration split, and wider, 120-degree LSA.

Equipped with the 113-degree cam, the blower engine produced 758 hp and 681 ft-lbs of torque. Replacing the 113-degree cam with the BTR cam increased the power output slightly to 768 hp and 679 ft-lbs of torque. The blower cam offered slightly more power at the top of the rev range, but the tighter LSA cam offered more low-speed torque.

An average (tight LSA) street cam works well, even on a blower engine, but is there extra power to be had from a wider LSA?

Tuning is critical on any LS3 application, but it's super critical on a supercharged combination. I employed a FAST management system to dial in each combination, but the air/fuel and timing curves were identical for each cam.

Effect of LSA on a Supercharged LSX (Horsepower)

113-Degree LSA Cam: 758 hp @ 6,200 rpm
120-Degree LSA Cam: 768 hp @ 6,200 rpm
Largest Gain: 10 hp @ 6,200 rpm

The wider LSA (typical of a positive displacement blower cam) improved power production above 5,400 rpm but offered a slight bump at 5,000 rpm as well. Down low, the wide LSA lost power to the tighter LSA, but low-speed power is usually not a problem with the immediate boost response offered by a positive displacement supercharger.

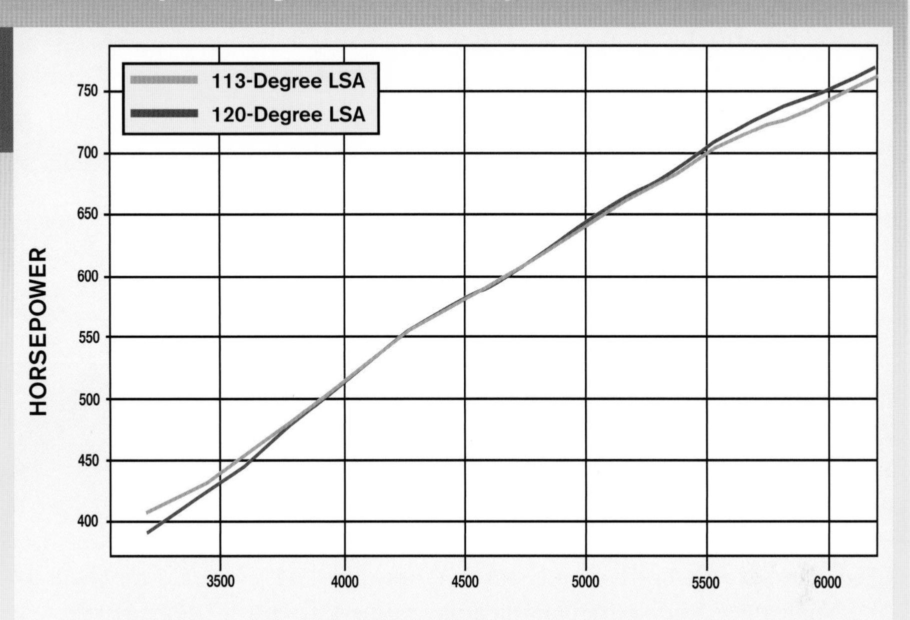

Effect of LSA on a Supercharged LSX (Torque)

113-Degree LSA Cam: 681 ft-lbs @ 4,200 rpm
120-Degree LSA Cam: 683 ft-lbs @ 4,300 rpm
Largest Gain: 19 ft-lbs @ 3,200 rpm

The 113-degree LSA cam increased torque very low in the rev range. From 3,200 to 3,900 rpm, the narrow LSA offered more low-speed torque, but lost out in terms of peak power to the wide LSA. The question now is, Where do you want your extra power?

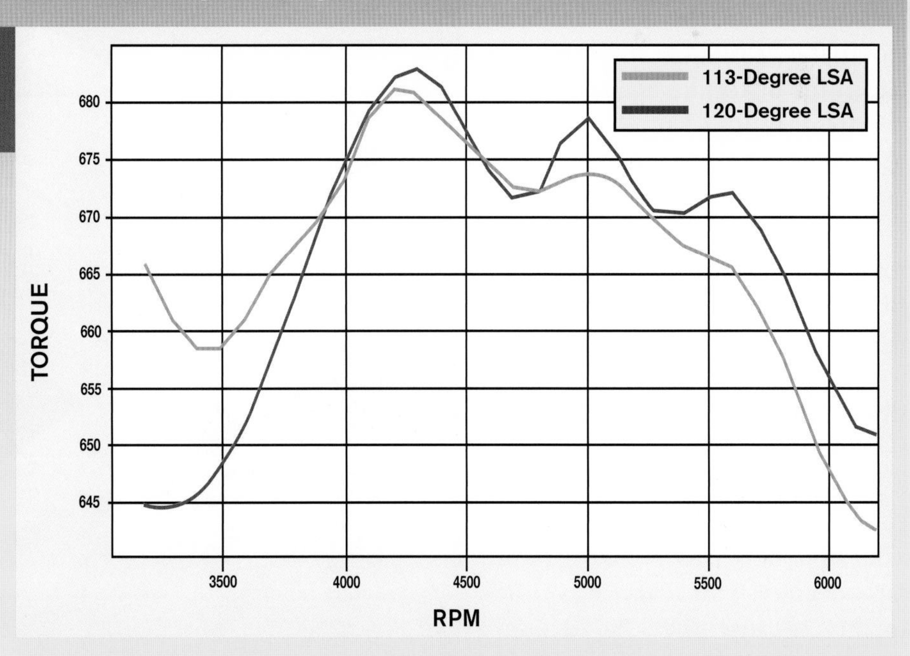

Test 6: Effect of LSA on a Stroker LS3

To find out how LSA affects the power curve on NA engines, I had Crane grind me a pair of cams with identical lift and duration values but altered LSAs. Both cams featured .624-inch lift (intake and exhaust) and 232/242-degree duration splits, but one cam featured a tight 108-degree LSA, while the other came in at 120 degrees. This obviously altered the cam timing events, but the test on LSA is interesting nonetheless.

The 402 stroker test engine was actually a hybrid of sorts, featuring an LS2 block and LS3 heads. The aluminum block was treated to a 4.0-inch stroker crank from Speedmaster, along with Carrillo rods and CP (flat-top) pistons. Topping off the hybrid stroker was a set of CNC-ported Chevy Performance L92 heads. Offering a tad more than 350 cfm, the heads flowed more than enough to support the intended power level for the test. The stroker also featured a FAST LSXR LS3 intake, Big Mouth throttle body, and 1⅞-inch Kooks headers.

JE supplied a set of asymmetrical, flat-top pistons for the 402 stroker.

The idea was to run the pair of cams to illustrate the power differences (if any) offered by the change in LSA. Equipped with the 120-degree cam, the stroker produced 570 hp at 6,200 rpm and 535 ft-lbs of torque at 5,200 rpm. After swapping to the 108-degree cam, the peak numbers stood at 572 hp at 6,300 rpm and 543 ft-lbs of torque at 5,100 rpm. A difference in 2 hp is not significant, but the real change came elsewhere in the curve.

The tighter LSA dramatically increased power production lower in the rev range. The cam swap netted an additional 36 ft-lbs of torque down low, but the additional torque gains continued through the rev range. Only for a short 150-rpm stint did the two cams produce the same power. The one downside to the tight (108-degree) LSA cam was idle quality because the idle vacuum was down significantly compared to the 120-degree cam.

The test engine was a 402-inch stroker LS2 block equipped with LS3 heads. The short-block included a 4340 Scat crank and 6.125-inch rods.

Effect of LSA on a Stroker LS3 (Horsepower)

120-Degree LSA Cam: 570 hp @ 6,200 rpm
108-Degree LSA Cam: 572 hp @ 6,300 rpm
Largest Gain: 10 hp @ 6,500 rpm

In terms of horsepower production, the tight 108-degree LSA cam offered more power everywhere. The 120-degree cam was able to equal the 108-degree cam for 200 rpm (from 6,000 to 6,200), but lost out everywhere else. I expected the 120-degree cam to come on strong at the top of the rev range, but on this stroker, the 108-degree was the better choice.

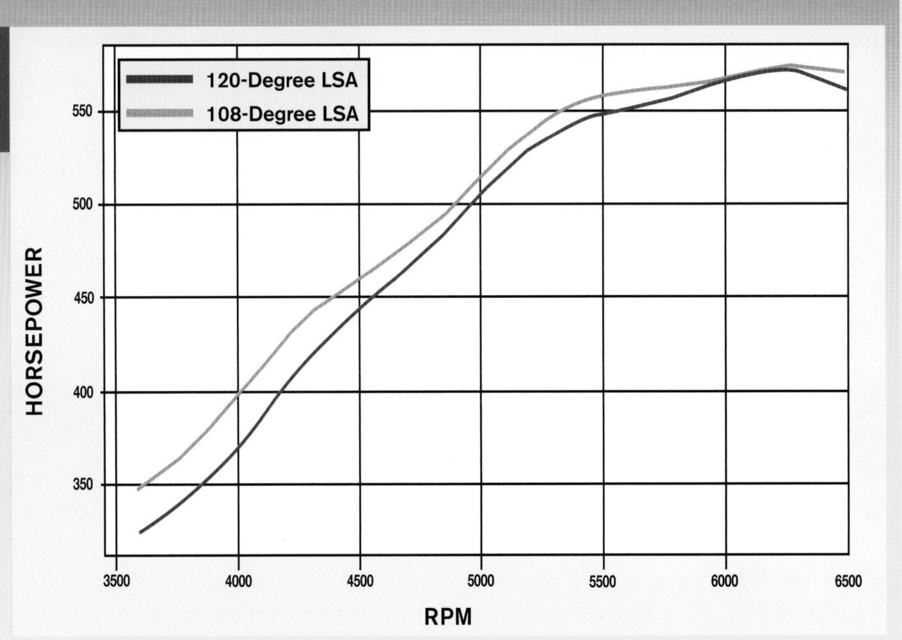

Effect of LSA on a Stroker LS3 (Torque)

120-Degree LSA Cam: 535 ft-lbs @ 5,200 rpm
108-Degree LSA Cam: 543 ft-lbs @ 5,100 rpm
Largest Gain: 36 ft-lbs @ 3,700 rpm

It is obvious from the torque curves that the tighter 108-degree LSA cam offered considerably more torque down low and through most of the curve. I saw this same scenario on the test with the supercharged application, although I did not try a 108-degree LSA on that engine. The downside to the tight LSA cam was idle quality, but if you are after power, the 108 was definitely the way to go.

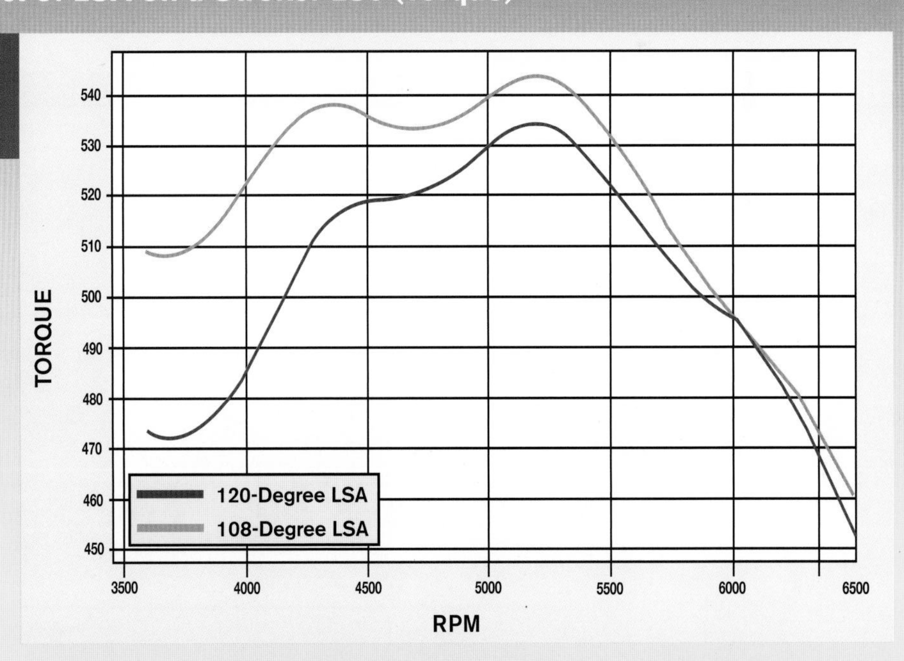

Test 7: Carb vs EFI Cam on a 417 LS3 Stroker

This test was interesting in that it came about after hearing a heated discussion online about the merits of cams designed specifically for carbureted LS applications. In truth, the carbureted cam design has less to do with what supplies the fuel than does the intake manifold design. Carbureted cams are generally designed for short-runner, single-plane manifolds rather than a specific carburetor. That enthusiasts even use the terms "carbureted" and "EFI cams" is reason enough to schedule a test.

As this and other tests revealed, the optimum cam for a given application works well with both forms of induction. By this I mean that if a cam offers more power in carbureted form, it will do so if you switch over to fuel injection.

The test engine for this comparison was a 417 stroker built from an LS3 aluminum block. Included were a 4.0-inch Scat crank, K1 6.125-inch rods, and JE forged pistons (with Total Seal rings). Topping the stroker was a set of GM Performance CNC L92 heads from Gandrud Chevrolet. The carbureted combination was run with an Edelbrock Victor Jr. intake and Holley 950 Ultra XP carburetor; the EFI combo included a FAST LSXR intake, Big Mouth throttle body, and 75-pound injectors.

In the carbureted corner was a Comp grind that offereda .623/.596-inch lift split, 247/258-degree duration split, and 110-degree LSA. The EFI cam was slightly milder with a .620/.596-inch lift split, 239/250-degree duration split, and wider 113-degree LSA. The graphs reveal that when tested with both the carbureted and EFI induction systems, the "carb" cam consistently offered more power. Neither the carb nor the EFI favored one of the cams; the carb cam was right for the application regardless of the induction system.

What cam works better, one designed for a carbureted application or one for an EFI application?

In addition to running both cams with an Edelbrock (carbureted) intake and Holley carb, I also ran the two cams with a long-runner, FAST EFI intake.

Carb vs EFI Cam on a 417 LS3 Stroker (Horsepower)

Carb Cam (carb combo): 628 hp @ 6,700 rpm
EFI Cam (carb combo): 611 hp @ 6,700 rpm
Largest Gain: 17 hp @ 6,700 rpm

In this comparison, I tested a pair of cams designed for carbureted and EFI applications. The carb cam was slightly wilder in specs and offered a tighter LSA. Run on a carbureted stroker, the carb cam offered more power above 5,300 rpm and below 4,700 rpm, but the two combinations yielded near identical torque values. Given the longer duration, I expected the carb cam to trade power down low to the EFI cam, but such was not the case.

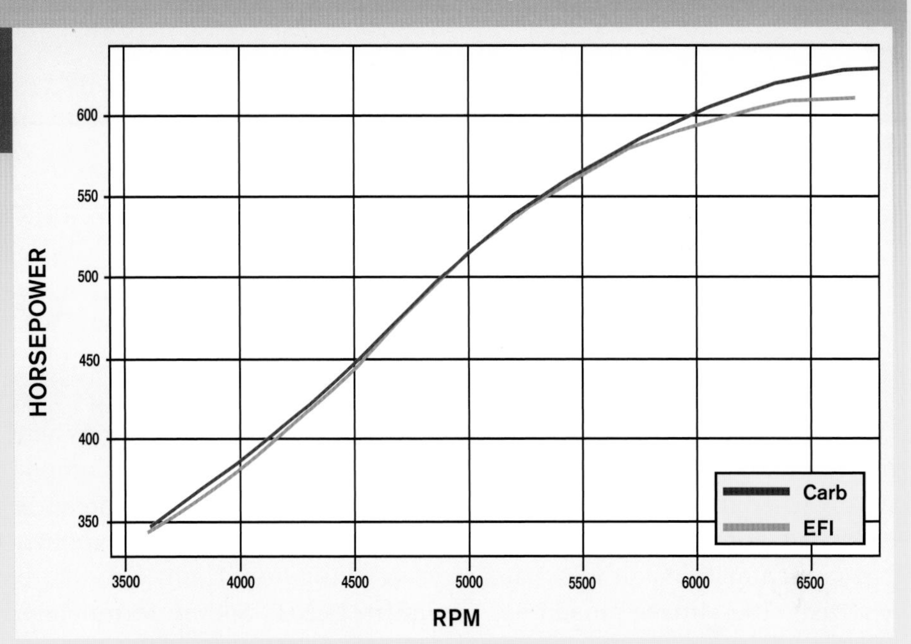

Carb vs EFI Cam on a 417 LS3 Stroker (Torque)

Carb Cam (EFI combo): 604 hp @ 6,400 rpm
EFI Cam (EFI combo): 585 hp @ 6,400 rpm
Largest Gain: 20 hp @ 6,500 rpm

I tested the carb versus EFI cam once again on the same stroker combination, but this time it was equipped with a long-runner, EFI manifold. Once again, the carb cam offered more power above 5,300 rpm. The two produced identical low-speed power (which surprised me), but the EFI cam offered slightly more power from 4,500 to 4,900 rpm.

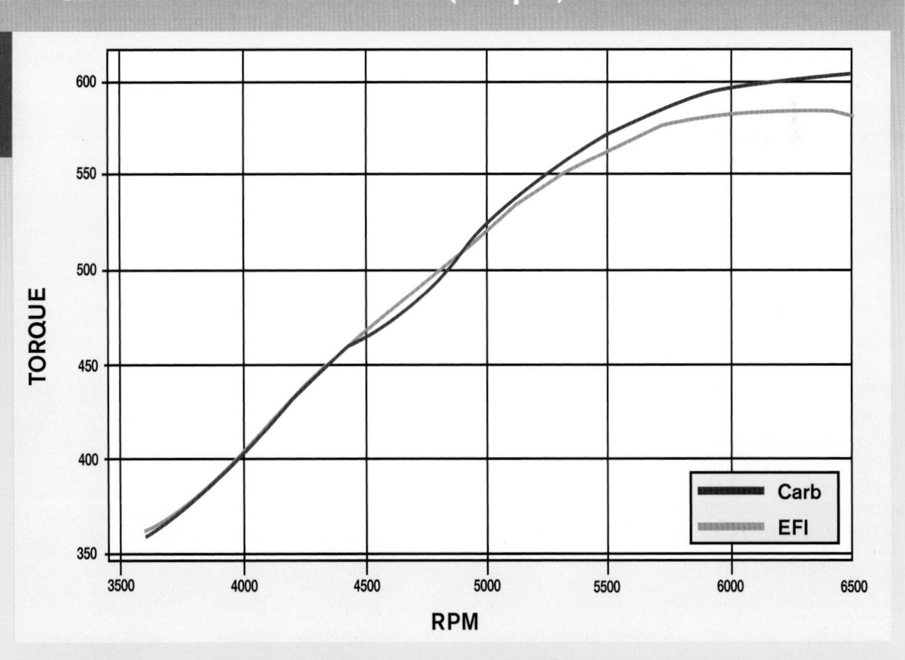

CHAPTER 4

HEADERS AND EXHAUST

Whether it's an LS3 or LS7, every engine responds to exhaust upgrades, especially when the upgrades include a dedicated long-tube header. In design and function, headers are much like the intake manifold, only in reverse. Like the intake, the exhaust manifolds (or tubular headers) are designed to not only allow, but actually improve, the exhaust flow out of the combustion chamber.

One of the common misconceptions about the exhaust system is that bigger is somehow better. When it comes to performance components, this bigger-is-better theme is quite common. It has been applied to everything from throttle bodies to cylinder heads, and unfortunately, this theme carries over to the exhaust systems. Common sense tells me that a set of 1⅞-inch (primary tube diameter) headers will outflow a set of 1¾-inch headers. Though it is obviously true that a larger header outflows a smaller set, does this mean that replacing the smaller headers results in a gain in power?

Shorty headers are a popular upgrade for LS engines, but just don't expect power gains on par with traditional, long-tube headers.

The answer to this question (like most of the dynamic equations involved with an internal combustion engine) is, it depends. You see, the power potential of a given set of headers has much more to do with their overall design (primary length, tubing diameter, and merge points) than the sheer size. Just like an intake manifold, the primary length plays a much more important role than the absolute airflow.

Contrary to popular belief, exhaust manifolds (or headers) are not actually designed to maximize flow. Were this true, maximizing performance would be a simple matter of building headers (and exhaust) with the largest diameter tubing you could find. The Zoomies on a Top Fuel dragster are a prime example of the "flow over scavenging" philosophy. Rather than just allowing the engine to exhale, proper header design actually evacuates the residual exhaust and even helps draw in the fresh induction charge, in effect helping to supercharge cylinder filling. This improved cylinder evacuation (and filling) happens by means of both kinetic energy and reflected pressure-wave scavenging.

Kinetic energy scavenging occurs by means of the release of pressure from the cylinder just as the exhaust valve opens. The elevated cylinder pressure (from the expansion created by the power stroke) finds the opening created by the recently opened exhaust valve (as the piston approaches bottom dead center, BDC). The compression wave created by this flow of the spent gases rapidly displaces the existing column of gas occupying the port. This compression wave increases pressure on the front (leading) side and reduces pressure on the back (trailing side) side. Because the compression wave travels through the port (primary tubing in a header) faster than the gas discharge speed (out of the cylinder from the upward moving piston), the low pressure on the trailing side of the compression wave actually helps draw out the remaining spent gases from the cylinder (in essence helping the piston do its job).

Long-tube headers are more effective than traditional cast-iron (or even short tubular) exhaust manifolds. While actual flow rates may be comparable between the two types of exhaust systems, long-tube headers improve power production in the same way long-runner intakes improve the volumetric efficiency on the intake side. Reflected pressure-wave scavenging occurs during the compression wave. That occurs when the exhaust valve opens to release the elevated cylinder pressure. And then the compression wave arrives at the end of the primary tube usually in the collector. Due to the increase in tubing diameter, the compression wave is allowed to expand and spread in all directions, and this is called rarefaction.

The depression created by this expansion causes air to rush in from the surrounding area, forcing the negative pressure wave back down the pipe to the awaiting exhaust port. This negative pressure wave helps further scavenge exhaust flow and aide induction flow into the cylinder during overlap. The length of the primary pipe has an effect on when (in RPM) this scavenging becomes most effective because the event should be timed so that the primary reflected wave arrives at the exhaust port when the piston has just passed top dead center (TDC). Since the reflected pressure waves travel at the speed of sound, the length of the primary pipes determine when the low-pressure wave coincides with the proper piston position, thus headers are actually tuned for particular engine speeds.

Although exhaust flow takes a back seat to scavenging, in some cases basic increases in flow become important. If available space, cost, or emissions compliance prohibit you from running a tuned header length, you can install a set of "shorty" headers. These were designed to replace the factory exhaust manifolds. Shorty headers offer little or no actual tuning (scavenging effect), but they can improve the flow rate over the factory manifolds. Shorty headers typically reduce vehicle weight because the tubular design weighs less than cast-iron exhaust manifolds.

On the plus side, shorty headers are much easier to install than traditional long-tube headers, which is very important for the do-it-yourselfer. The downside is that because they have insufficient primary length to provide adequate scavenging, the power gains offered by shorty headers are considerably less than those offered by long-tube headers. The gains offered by shorty headers increase with the power output of the test engine; just don't expect the 15- to 20-hp gains you often see with long-tubes.

Are a set of 1⅞-inch headers really better than a set of 1¾-inch headers for your LS3? See page 64.

Collector length plays a part in the shape of the power curve. Contrary to popular belief, open headers are not the hot setup, even for race cars.

Supercharged applications (such as the Kenne Bell) adapt well to different header configurations.

Test 1: Stock Exhaust Manifolds vs Shorty Headers on an LS3

If you own an LS3 or LS7, before you purchase any performance part, you should be asking one very important question. Does it make more power? After all, why go to all the expense and trouble to purchase and install the upgrade if it doesn't add any power? For this test, this question was applied to headers, or more specifically, shorty headers.

You see elsewhere in this chapter that long-tube headers offer substantial power gains thanks to exhaust scavenging, but shorty headers operate by (theoretically) increasing flow. The problem with any performance question is that it needs to be specific. The question isn't so much do shorty headers work (and they do), but rather how well do they work on your exact combination? Obviously, a test run on a stock LS3 doesn't illustrate how well they work on a more powerful combo. Unfortunately, it is impossible to run tests on every combo, but I can make some educated guesses after having run similar tests on other combos.

To prepare for this header test, I first needed two very important components: headers and a test engine. I started the test with a set of factory, cast-iron exhaust manifolds, but I compared them to a set of shorty headers. Both were fitted with a collector extension to simulate the exhaust system. The collector extension is important, especially on long-tube headers (see Tests 5 and 7).

The test engine was the trusty LS3 crate engine from Gandrud Chevrolet. In fact, this was one of the very first tests I ran after taking delivery of the engine. The LS3 was configured for dyno use with a FAST manual throttle body, Holley Dominator EFI system, and Meziere electric water pump.

Run with stock manifolds, the LS3 produced peak numbers of 495 hp and 490 ft-lbs of torque. After installing the shorty headers, the peaks jumped to 497 hp and 492 ft-lbs. The largest power gains offered were 5 hp and 3 ft-lbs, but know that on more powerful combinations, the shorty header can be worth even more power (I tested them previously on a stock and modified 5.3). Just don't expect the kind of gains you get from long-tube headers.

The air/fuel and timing curves were dialed in using a Holley Dominator EFI system.

The shorty header (not installed) test was run on a bone-stock LS3 crate engine from Gandrud Chevrolet.

Stock Exhaust Manifolds vs Shorty Headers on an LS3 (Horsepower)

LS3 Stock Exhaust Manifolds: 495 hp @ 5,700 rpm
LS3 Shorty Headers: 497 hp @ 5,700 rpm
Largest Gain: 5 hp @ 6,500 rpm

The shorty headers were worth very little power over the stock exhaust manifolds on this stock LS3. Had I tested at a higher power level, the headers may have shown slightly greater gains, but the shorty headers lack true scavenging offered by long-tube headers.

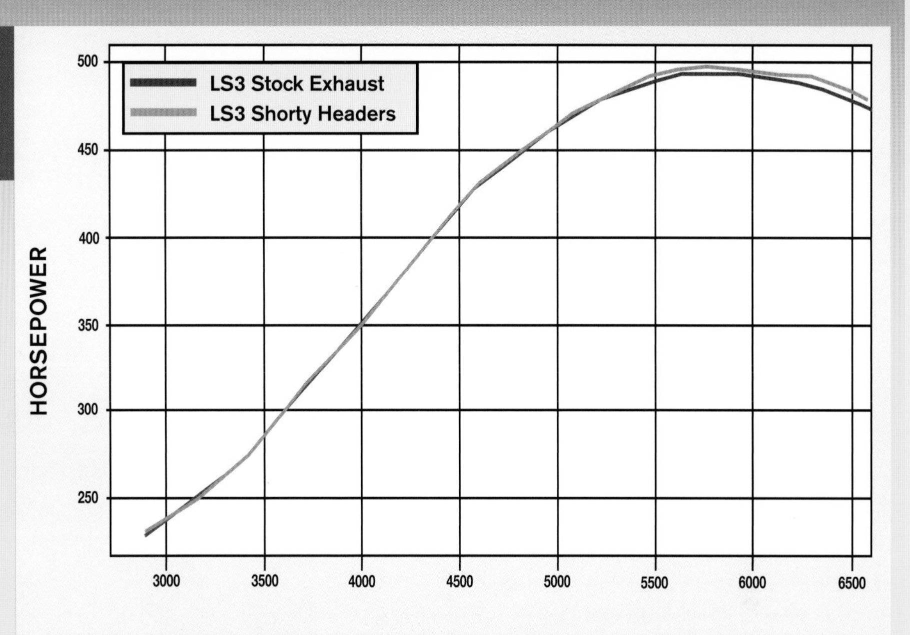

Stock Exhaust Manifolds vs Shorty Headers on an LS3 (Torque)

LS3 Stock Exhaust Manifolds: 490 ft-lbs @ 4,700 rpm
LS3 Shorty Headers: 492 ft-lbs @ 4,600 rpm
Largest Gain: 3 ft-lbs @ 4,600 rpm

With torque gains of just 3 ft-lbs, it is hard to justify replacing the stock manifolds, but the gains should increase with other mods. It should also be pointed out that the shorty headers weigh significantly less than the heavy stock manifolds and the swap improves the power-to-weight ratio, which can improve acceleration.

I ran this same test on a stock and modified 5.3L, and the gains offered by shorty headers improved after the modifications.

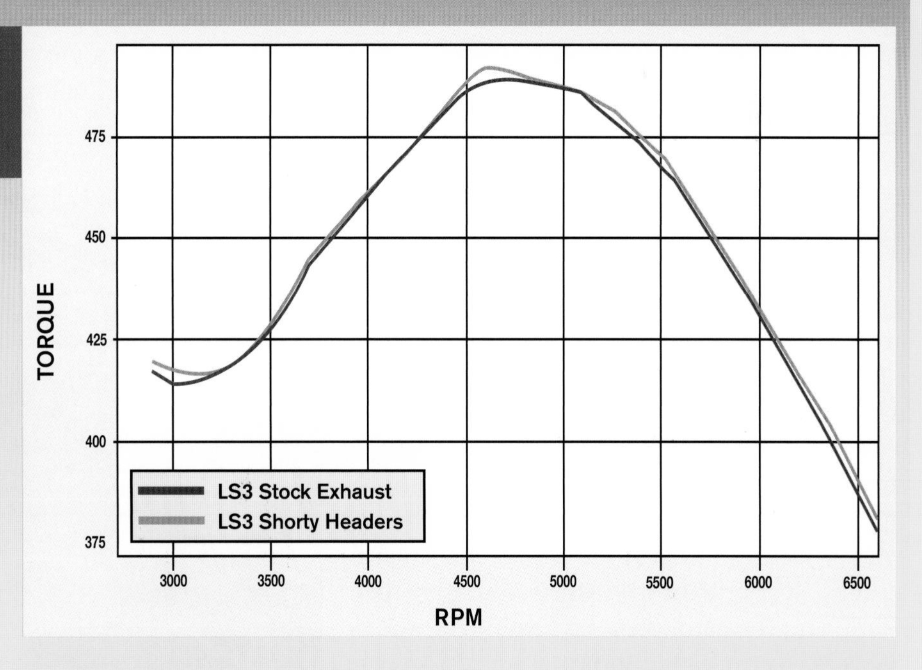

Test 2: Shorty vs Long-Tubes on a Modified LS3

Long-tube headers are one of the least understood aspects of any exhaust system. The confusion stems from the fact that, unlike the remainder of the exhaust, where flow is the major concern, headers provide a tuning effect on the power band much like an intake manifold. Some enthusiasts (and tuners alike) seem to be under the misconception that headers add power by offering more flow than a stock exhaust manifold. In truth, the power gains offered by true long-tube headers come not from flow, but rather from the scavenging effect offered by the primary (and collector) length.

Much like the intake manifold, the tubular header improves the power output by increasing the intake flow into the combustion chamber. This is accomplished by scavenging the exhaust through both pressure-wave and reflected-wave tuning. The reflected wave occurs when the high-pressure exhaust gas exits the primary tubing into the collector of a 4-into-1 header (or secondary Y-section of the Tri-Y design). The compression wave is allowed to expand and spread out in all directions. This expansion creates a depression (rarefaction) and this negative pressure wave travels back up the primary tube to the exhaust valve. If timed properly (with proper design), this negative pressure wave helps scavenge exhaust from the combustion chamber and aids induction during the overlap period (when both intake and exhaust valves are open).

To illustrate the power gains offered by long-tube headers, I put them to the test on an LS3 crate engine supplied by Gandrud Chevrolet. For this header test, the LS3 was upgraded with a Comp 459 cam that offered a.617/.624-inch lift split, 231/239-degree duration split, and 113-degree LSA. Additional mods included a manual FAST throttle body, FAST injectors, and a Holley HP management system. Ever present were the Aeromotive fuel pump and regulator, Lucas synthetic oil, and a Meziere electric water pump.

Run on the dyno with the shorty headers feeding collector extensions, the cam-only LS3 produced 543 hp and 505 ft-lbs of torque. After installation of the Hooker long-tube headers (with collector extensions), the power output jumped to 558 hp and 513 ft-lbs of torque. Had the headers simply improved flow, the gains would be most prevalent at the top of the rev range, but the tuning effect offered torque gains through the entire rev range.

The engine for this shorty versus long-tube header test was a crate LS3 equipped with a powerful (but streetable) Comp cam. Even with shorty headers, the combo exceeded 540 hp.

The test was run with the factory LS3 intake manifold, FAST injectors, and a FAST 92-mm (manual) throttle body.

Shorty vs Long-Tubes on a Modified LS3 (Horsepower)

Shorty Headers: 543 hp @ 6,200 rpm
Long-Tube Headers: 558 hp @ 6,300 rpm
Largest Gain: 18 hp @ 6,500 rpm

Contrary to popular belief, headers do not simply flow better than stock exhaust manifolds or (in this test) shorty headers. All things being equal, shorter tubes flow more than the same-size longer tube, so the power gains have much more to do with the resonance tuning offered by the long-tube headers than a simple increase in flow. Long-tube headers offer significantly more power than shorty headers. Tested on this cam-only LS3, the long-tube headers increased the peak power output from 543 hp to 558 hp, with significant gains through the entire rev range.

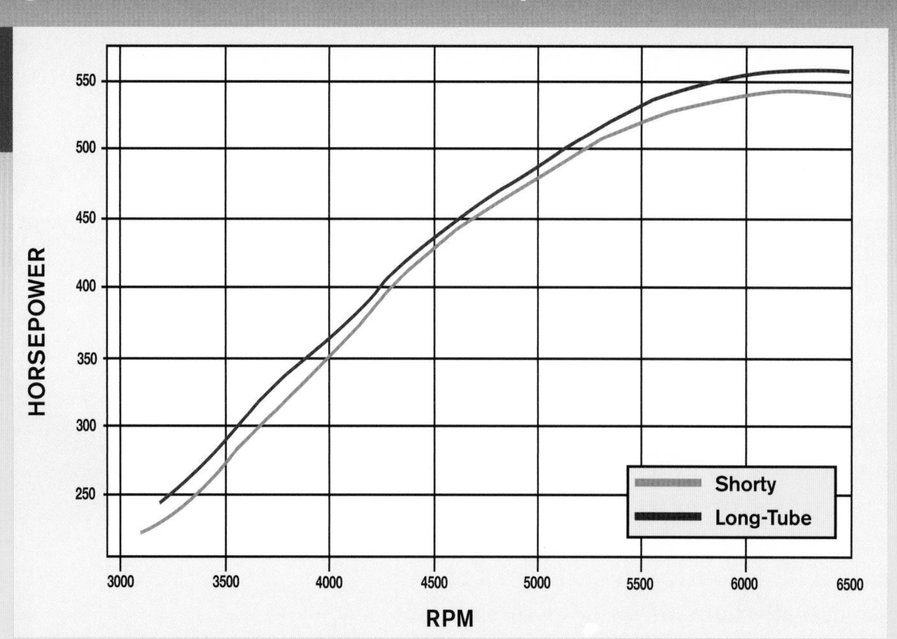

Shorty vs Long-Tubes on a Modified LS3 (Torque)

Shorty Headers: 505 ft-lbs @ 4,700 rpm
Long-Tube Headers: 513 ft-lbs @ 4,700 rpm
Largest Gain: 28 ft-lbs @ 3,600 rpm

The torque gains offered by the long-tube headers were even more impressive because the long tubes increased torque production by as much as 28 ft-lbs at 3,600 rpm. In fact, the scavenging offered by the headers improved low-speed torque even more than peak power. Such is the benefit of resonance tuning. Torque gains are always welcome, but even more so when you improve torque through the entire rev range.

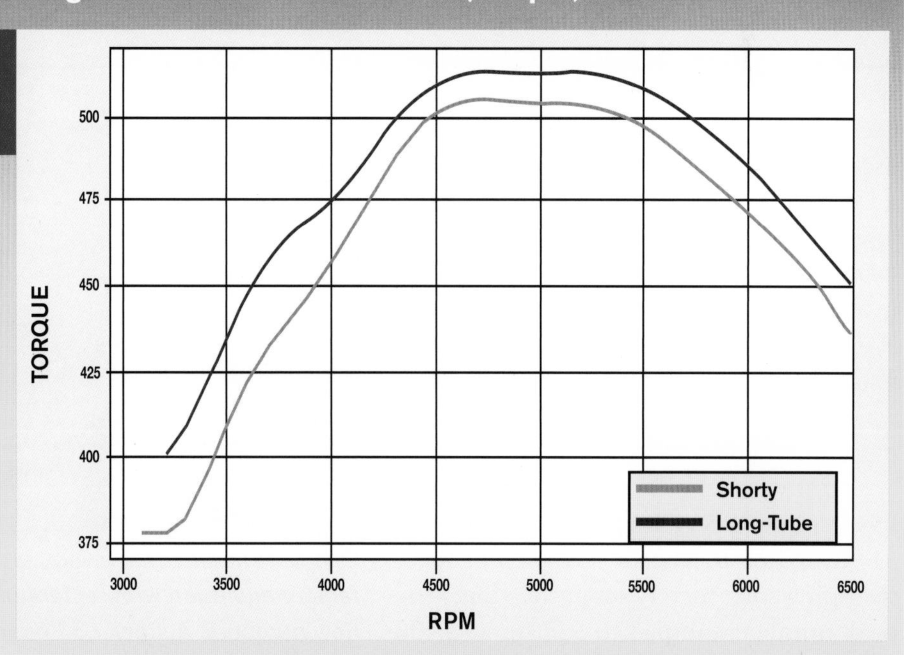

Test 3: 1¾-inch vs 1⅞-inch on a Mild LS3

When it comes to increasing performance, there is one steadfast rule: All the good air that goes into making horsepower must eventually find its way out. Thus, adding cool camshafts, intake manifolds, and even superchargers to help improve the airflow into the engine will be useless (okay, maybe not useless but certainly much less effective) if the engine is not able to rid itself of the exhaust. Cork up a serious LS performance engine and watch it struggle and gag on its own exhaust fumes.

Adding the right header can significantly improve exhaust flow and add power through a process called scavenging. You see, a header is much more than a simple set of tubes welded together to direct the exhaust flow. A true header provides a direction for the exhaust and can also actually help draw the spent gases out of the combustion chamber. The effectiveness of scavenging is determined by a number of design criteria, including primary tubing length and diameter, collector size, and even the overall design.

This test was designed to illustrate the difference in power curves offered by 1¾- and 1⅞-inch primary tubing. It was run on a modified LS3 crate engine (once again from GM Performance, courtesy of Gandrud Chevrolet). The same engine was used in Test 2, but the LS3 was fortified with a Comp 259 cam, FAST throttle body, and injectors. The test ran a set of 1¾-inch QTP headers and then a set of larger 1⅞-inch headers from American Racing Headers. As is evident by the results, the difference in manufacturers (with the same size tubing) was less than the difference between primary tubing diameters.

Run with the 1¾-inch headers, the LS3 produced 554 hp at 6,300 rpm and 514 ft-lbs of torque at 4,800 rpm. After swapping on the 1⅞-inch headers, the peak numbers jumped to 558 hp at 6,400 rpm, but the maximum torque remained at 514 ft-lbs (at a higher 5,300 rpm). The larger headers offered more peak power, but also offered more low-speed torque. The 1¾-inch headers showed well in the mid-range, but the largest difference in power or torque was just 4 to 5 hp.

The most significant upgrade applied to the LS3 crate engine was the installation of a Comp 459. This Comp cam offered a.617/.624-inch lift split, 231/239-degree duration split, and 113-degree LSA.

One reason that cam swaps are so effective on the LS3 is that they are factory equipped with rectangular-port cylinder heads that offer huge flow numbers. A stock LS3 head flows as much as 318 cfm, enough to support nearly 650 hp.

1¾-inch vs 1⅞-inch on a Mild LS3 (Horsepower)

1¾-Inch Headers: 554 hp @ 6,300 rpm
1⅞-Inch Headers: 558 hp @ 6,400 rpm
Largest Gain: 5 hp @ 6,500 rpm

There are two ways to look at the results of this test. The first is to dismiss the minor differences between the 1¾- and 1⅞-inch headers as insignificant; after all, the difference of 4 to 5 hp on an engine that exceeded 550 hp represents just 1 percent. Add to that the smaller header that actually made more power in the midrange and the choice becomes more difficult. View number two says that any amount of extra power is welcome and, at this power level, you have to fight for every gain possible. Those 4- to 5-hp gains eventually add up to something special, and often make the difference between a good combination and a winning one.

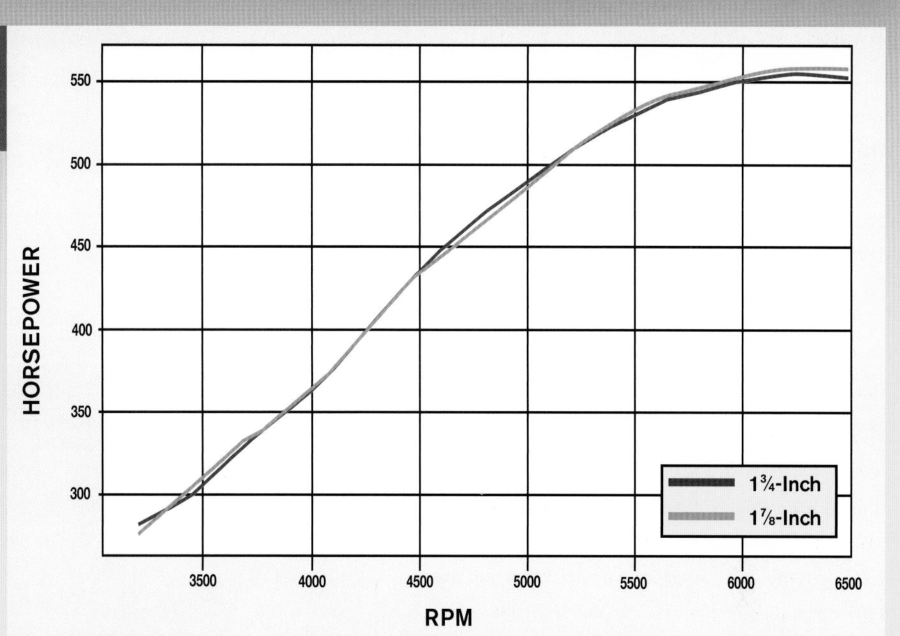

1¾-inch vs 1⅞-inch on a Mild LS3 (Torque)

1¾-Inch Headers: 514 ft-lbs @ 4,800 rpm
1⅞-Inch Headers: 514 ft-lbs @ 5,300 rpm
Largest Gain: 5 ft-lbs @ 4,700 rpm

The interesting thing about this test was the fact that the 1⅞-inch long-tube headers actually made more low-speed torque than the smaller-diameter headers. In the mid-range, the smaller headers took over, but above 5,250 rpm, the larger-diameter headers pulled away once again. This trade-off in power production is what makes header design (and selection) so difficult. Header design is further limited by fitment, but even if you had free rein, it would be impossible to make a universal header for every application.

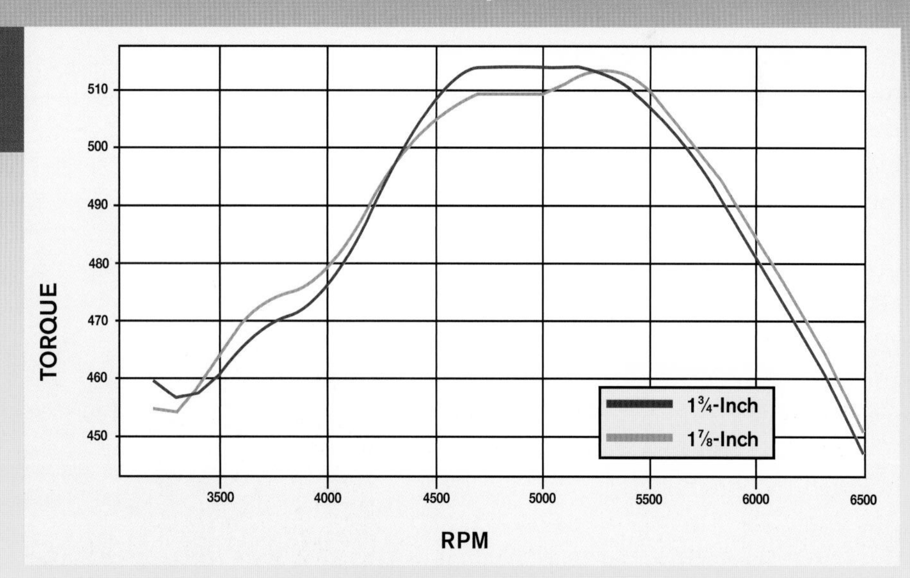

Test 4: 1¾ and 1⅞ vs 1⅞ Steps on an LS7

The LS7 test engine began life as a simple LS3. The aluminum block received Darton sleeves that allowed a complete Manley stroker assembly that included 4.130-bore flat-top pistons, H-beam rods, and a Platinum-series (4.0-inch) stroker crank. Sealing the pistons was a Total Seal ring package. The custom LS7 cam featured a .644-inch lift, a 246/254-degree duration split (at .050), and was teamed with an adjustable cam sprocket and Aviad dry-sump oiling system. Topping the big-bore LS7 was a set of CNC-ported CMP Brodix SI LS7 heads.

These custom LS7 castings offered 258-cc intake ports, 108-cc exhaust ports, and 62.5-cc combustion chambers. Head flow peaked at just over 395 cfm on the intake and 259 cfm on the exhaust, allowing them to support more than 800 hp on the right application. This modified LS7 was run on the dyno with Lucas synthetic oil, a Holley Dominator management system, and two different styles of headers. Also present on the impressive stroker was an ATI dampener, Meziere electric water pump, and FAST 102-mm throttle body.

This test run on the modified LS7 involved a comparison between a set of 1¾ and 1⅞-inch step headers and a set of fixed-diameter 1⅞-inch headers. The two headers from American Racing Headers and Kooks offered similar primary lengths and identical collector lengths and diameters. This test illustrated that some LS combinations are considerably less sensitive to changes in header design.

Equipped with the step headers, the modified LS7 produced 667 hp at 6,800 rpm and 581 ft-lbs of torque at 5,000 rpm. After replacing the step headers with straight 1⅞-inch primaries, the engine produced 666 hp at the same 6,800 rpm and 580 ft-lbs at an identical 5,000 rpm. Basically, the engine produced not only the same peak horsepower and torque values, but also the same overall power curves. The difference in power between the two headers was never greater than 1 to 2 hp, or the amount you might see in a back-to-back test with the same headers. On this combination, the two headers were identical.

The 427 LS7 test engine actually started out life as an LS3. The aluminum LS3 block received Darton sleeves to allow the bore to be punched out to 4.130 then combined with a 4.0-inch stroke crank.

The LS3/LS7 hybrid featured a set of CNC-ported CMP/Brodix LS7-based cylinder heads, MSD Atomic intake, and FAST throttle body.

1¾ and 1⅞ vs 1⅞ Steps on an LS7 (Horsepower)

Step Header: 667 hp @ 6,800 rpm
1⅞-inch Header: 666 hp @ 6,800 rpm
Largest Gain: 1 hp @ 6,800 rpm

No, there is nothing wrong with the graphs for this test; the power outputs of the two header styles were nearly identical. Despite what many would consider to be a major change in header design (stepping from 1¾ to 1⅞ inches), the two headers differed by a scant 1 hp (a statistically insignificant amount). Some combinations are simply less sensitive to changes in header design.

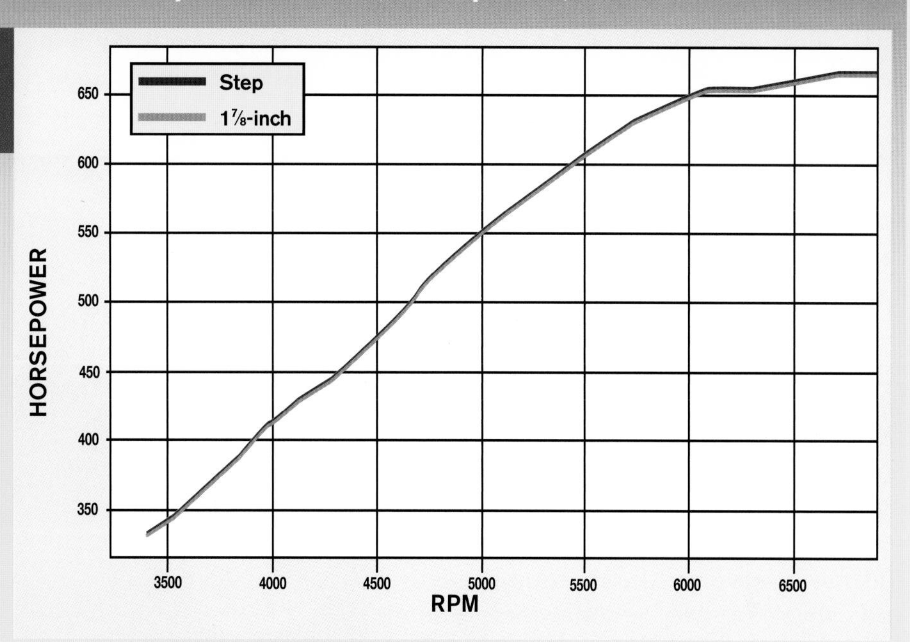

1¾ and 1⅞ vs 1⅞ Steps on an LS7 (Torque)

Step Header: 581 ft-lbs @ 5,000 rpm
1⅞-inch Header: 580 ft-lbs @ 5,000 rpm
Largest Gain: 1 ft-lb @ 5,000 rpm

The scale on the torque curve allowed subtle changes to be illustrated more clearly, but I am still talking about changes of 1 to 2 ft-lbs. The differences in torque production would be impossible to distinguish on the street or track because the two produced peak torque values within 1 ft-lb.

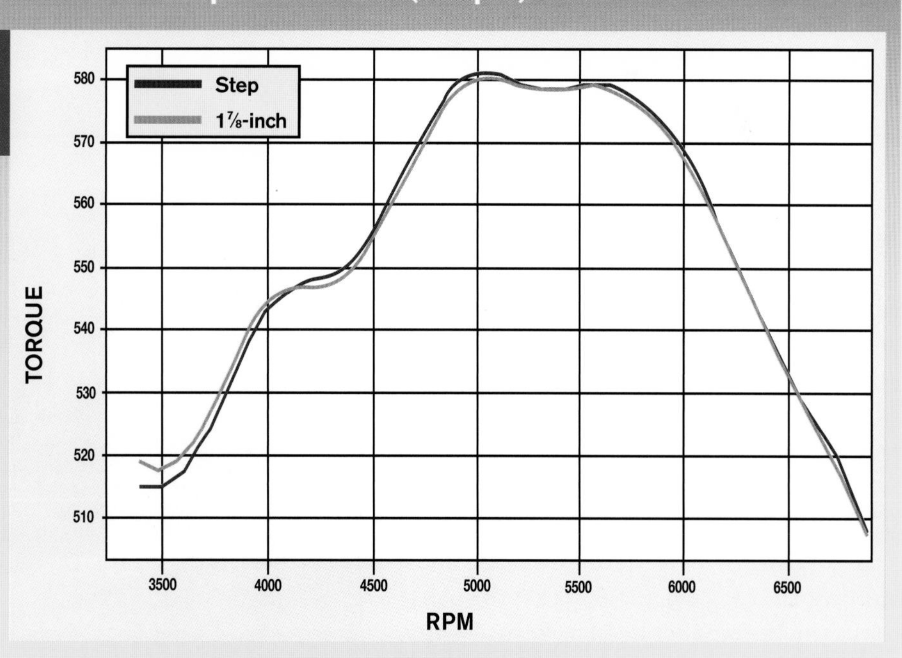

Test 5: 1¾ vs 1⅞ SC B15 LSX

If you review the results of Test 3, you see that replacing 1¾-inch headers with larger 1⅞-inch headers was of little value on the cam-only LS3, but what about on a wilder, supercharged LSX combination? It stands to reason that a high-horsepower, supercharged combination would require larger headers, but remember, headers are more than just exhaust flow. The proper header design seeks to improve intake flow through scavenging by creating a negative pressure wave in the chamber.

Although the presence of boost changes this test (compared to the mild LS3 in Test 3), there is another major change. The Whipple supercharger system also employed a short-runner intake manifold. As was evident in Chapter 1, the runner length has a tuning effect on the power curve. The intake runner length and exhaust primary length work together to optimize cylinder filling, but when you change one of them (for instance, shortening intake runner length), it throws off the delicate balance. Does this mean the short-runner intake is less sensitive to primary diameter? That's why you test!

This test ran a set of 1¾-inch headers against a set of 1⅞-inch headers on a supercharged B15 crate engine from General Motors. The B15 was designed specifically for boost with a forged crank and pistons. Also present were (as-cast) LSX/LS3 heads installed with ARP head studs, a Comp 469 cam (.617/.624 lift split, 231/247 duration split, 113 LSA), and a Whipple 4.0 supercharger. The supercharged combo was tuned with a FAST XFI management system using 150-pound Holley injectors.

Run with the smaller 1¾-inch headers, the supercharged combo produced 758 hp and 679 ft-lbs of torque at a peak boost reading of 14.7 psi (at 6,200 rpm). After replacing the smaller headers with 1⅞-inch headers, the peak power numbers jumped to 774 hp and 684 ft-lbs of torque. The power gains increased with engine speed and showed signs of increasing even further had I elected to rev the engine higher; the power and boost were both climbing rapidly. Run on the high-horsepower, supercharged LSX, the larger-diameter tubing was definitely the hot setup.

The B15 LSX crate engine was set up for boost with forged internals and these six-bolt LS3 LSX heads.

It was necessary to upgrade the valvesprings when I replaced the factory LS9 cam with the Comp 469 unit. The stock springs were replaced with a set of dual springs from BTR.

1¾ vs 1⅞ SC B15 LSX (Horsepower)

1¾-inch Headers: 758 hp @ 6,200 rpm
1⅞-inch Headers: 774 hp @ 6,200 rpm
Largest Gain: 16 hp @ 6,200 rpm

The horsepower gains offered by the larger headers increased with engine speed. Had I elected to run this supercharged B15 crate engine higher in boost and engine speed, the gains offered by the header swap would increase. The supercharged LS3 responded well to the larger primary tubing.

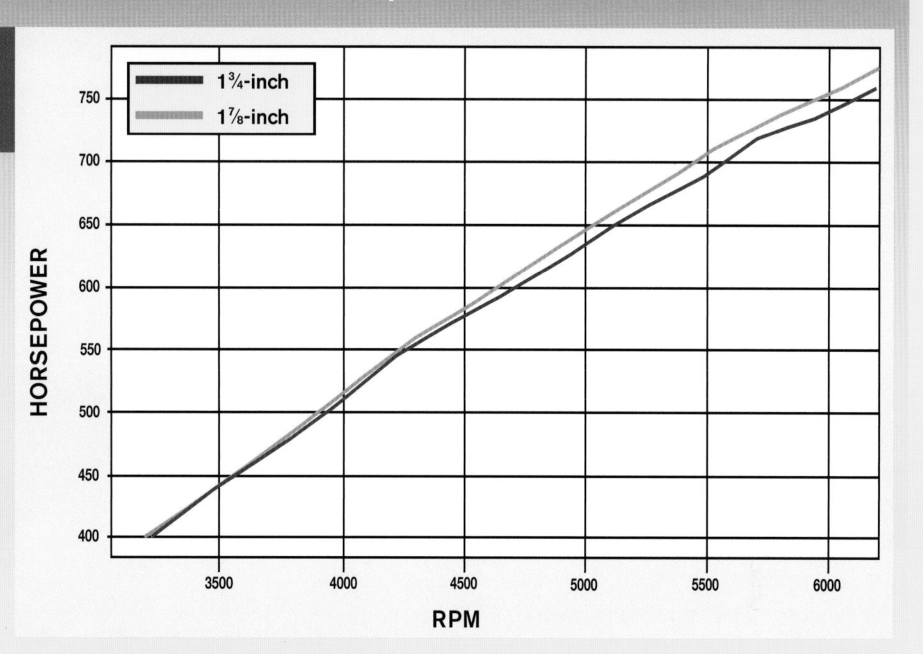

1¾ vs 1⅞ SC B15 LSX (Torque)

1¾-inch Headers: 679 ft-lbs @ 4,200 rpm
1⅞-inch Headers: 684 ft-lbs @ 4,300 rpm
Largest Gain: 13 ft-lbs @ 4,900 rpm

The torque gains offered by the larger-diameter headers started at 3,600 rpm but really got going past 4,200 rpm. Run to 6,000 rpm, the torque gains were pretty consistent and substantial enough to merit use of the larger headers. Note the slight dip in torque production below 3,500 rpm, where the smaller headers offered increased torque production.

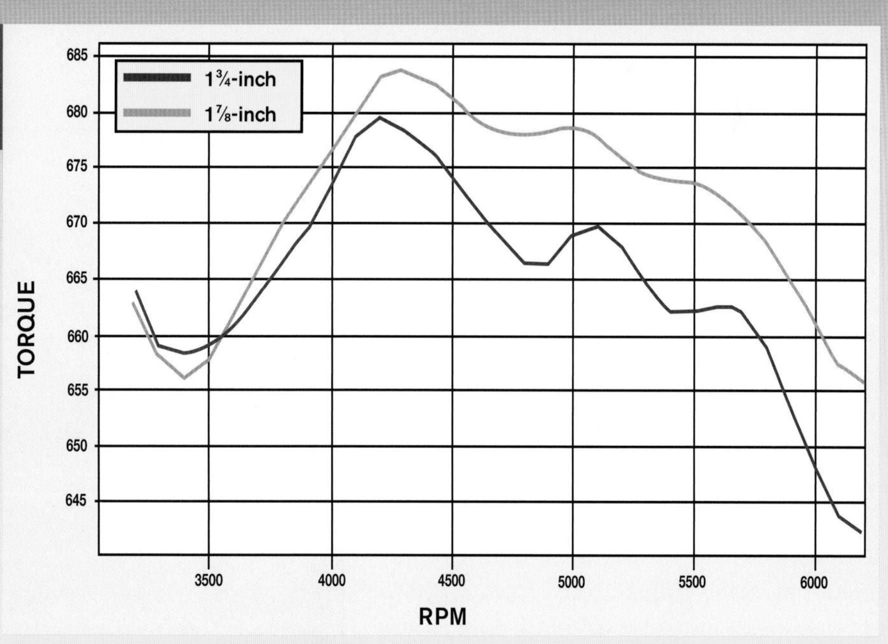

Test 6: Effect of Collector Length on a 6.0 LS3 Hybrid

This test on a mild test engine displaced just 6.0 liters. The 6.0 was supplied by Blueprint Engines. The GM version (PN 364LS) supplied by BPE came as a long-block, meaning it lacked a few important components required to make it run. The BPE 364 long-block included a number of desirable features, including the 6.0 iron block, forged pistons, and a static compression ratio of 9.8:1.

The factory GM (3.622-inch stroke) crank was configured with a 24X crank trigger and the block was equipped with a rear cam sensor. Topping the 6.0 short-block was a set of BPE LS3-based aluminum heads (their own castings). The heads (PS8015) were configured with 259-cc intake ports, 102-cc exhaust ports, and 72-cc combustion chambers. Helping flow was a set of 2.165-inch intake valves and 1.60-inch exhaust valves set at 12-degree valve angles. Working with the rectangular-port heads was a single-pattern performance cam that featured a .556-inch lift, 223-degree duration, and 114-degree LSA. The final touches were an adjustable timing chain and oil pump.

To complete this long-block, I installed the Holley dual-plane intake, 950 Ultra XP carburetor, and MSD ignition controller. The test ran a set of Hooker 1⅞-inch swap headers with two collector lengths (16 and 42 inches).

Run with the shorter collector extension, the BPE 6.0 produced 494 hp at 6,200 rpm and 457 ft-lbs of torque at 4,400 rpm. After installation of the longer collector extensions, the BPE 364 produced the same 494 hp, but peak torque decreased slightly to 450 ft-lbs at 4,500 rpm. The longer collectors improved torque production below 3,600 rpm, but power suffered from 3,700 to 5,000 rpm. From there to 6,500 rpm, there was little difference, though the longer collector length did start to fall off ever so slightly at the top of the rev range. Because cam timing initiates the all-important reflected waves, the cam timing must work well with the intake and exhaust components for maximum power production.

Supplied by Blueprint Engines (BPE), the test engine was an iron-block 6.0 equipped with a mild cam and BPE LS3 heads.

The 6.0 hybrid was equipped with a Holley induction system that included a dual-plane intake and 950 Ultra XP carburetor.

Effect of Collector Length on a 6.0 LS3 Hybrid (Horsepower)

Short Extension: 494 hp @ 6,200 rpm
Long Extension: 494 hp @ 6,200 rpm
Largest Gain: 6 hp @ 4,400 rpm

Run on this carbureted 6.0 equipped with LS3 heads, the longer collector length improved power production up to 3,600 rpm, but power dropped off up to 4,900 rpm. From 5,000 rpm up, the collector length had little effect on power production. This mild, carbureted 6.0 did not respond well to the change in length.

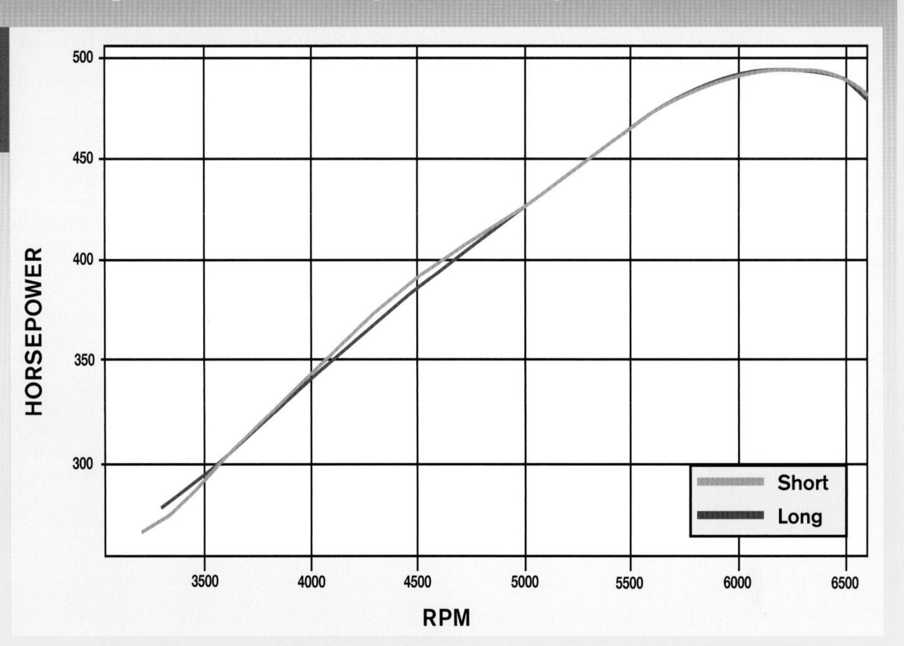

Effect of Collector Length on a 6.0 LS3 Hybrid (Torque)

Short Extension: 457 ft-lb @ 4,400 rpm
Long Extension: 450 ft-lbs @ 4,500 rpm
Largest Gain: 9 ft-lbs @ 3,300 rpm

Changes in the torque curve offered by the collector length were most prevalent below 5,000 rpm. At 3,300 rpm, the longer collector length increased torque production by 9 ft-lbs on this 6.0 LS3 hybrid, but torque suffered above that point.

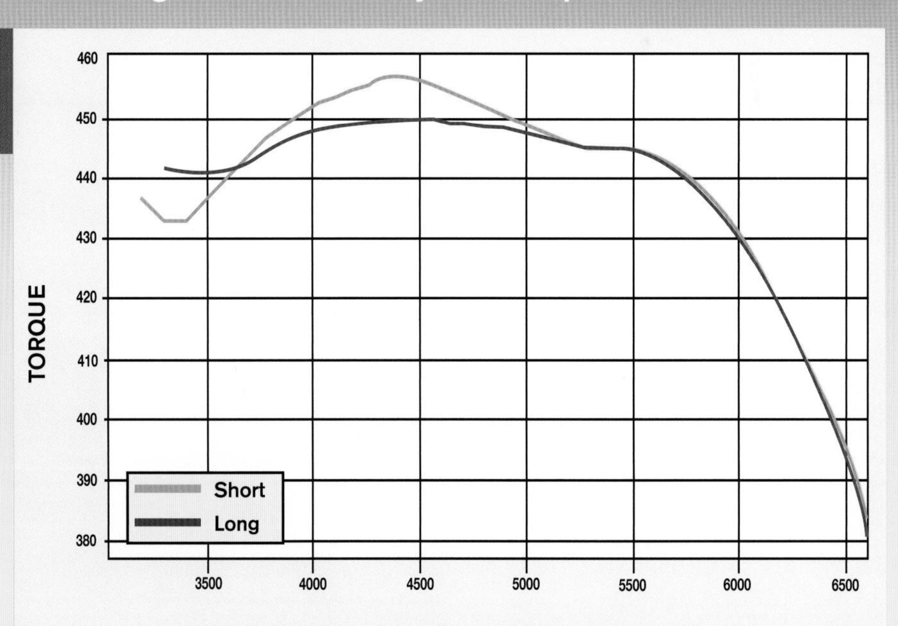

Test 7: Shorty vs 1⅞ Hooker KB SC LS3

To run a supercharged header test, I needed two things: a suitable test engine and a supercharger. The LS3 crate engine was supplied by my good friends at Gandrud Chevrolet and the twin-screw supercharger came courtesy of Kenne Bell.

Plucked right off the assembly line, the LS3 was essentially a stock 6.2 Corvette engine. For my needs, the important part was that it featured the impressive LS3, rectangular-port aluminum heads. For dyno use, I replaced the factory Corvette dampener (crank pulley) with a Camaro/Truck offset, then installed a Meziere electric water pump.

I owe a huge shout-out to Kory at Turnkey Engine Supply and Mike at Kenne Bell, as the two wrangled up a suitable drive system that allowed me to run the twin-screw supercharger on the dyno sans accessories.

Now all I had to do was swap a couple of headers on the dyno, but first I had to make sure that the supercharged combination was repeatable. Dialing the tune-in was a Holley HP management system controlling a set of FAST 75-pound injectors.

For this test, I relied on a 2.8 Kenne Bell supercharger. The unit was capable of supporting in excess of 1,000 hp, but because I was dealing with an LS3 equipped with stock internals, I kept the boost pressure to a reasonable level. The supercharger was set up with a 3.75-inch blower pulley that resulted in a peak boost pressure of 9.9 psi at 6,500 rpm. Thanks to plenty of belt wrap, the new six-rib drive assembly ran flawlessly during testing.

First up were the shorty headers, which produced peak numbers of 706 hp at 6,400 rpm and 632 ft-lbs of torque at 4,400 rpm. After installation of the Hooker 1⅞-inch long-tube headers, the peak numbers jumped to 724 hp and 645 ft-lbs of torque. The boost pressure actually dropped by a couple of tenths of a pint after installation of the long-tube headers, but not nearly as much as I expected given the power difference.

This test is yet more evidence that long-tube headers offer significant power gains on both naturally aspirated and supercharged combinations.

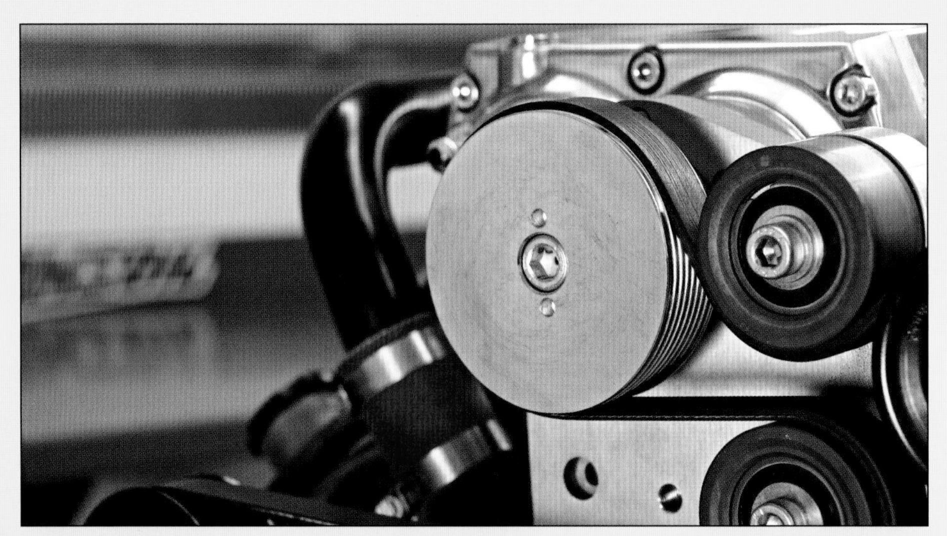

Boost supplied by the supercharger can be altered by simply swapping the blower (and/or crank) pulley. The pulley was not changed during the header test.

The LS3 crate engine was upgraded with a Kenne Bell 2.8 twin-screw supercharger kit. Capable of supporting 1,000 hp, the 2.9 was more than adequate for the mild LS3.

Shorty vs 1⅞ Hooker KB SC LS3 (Horsepower)

Shorty Headers: 706 hp @ 6,400 rpm
Hooker 1⅞ Headers: 724 hp @ 6,400 rpm
Largest Gain: 21 hp @ 5,700 rpm

Changes in the torque curve offered by the long-tube headers were most prevalent above 4,500 rpm, but it's hard to argue with an extra 10 to 12 ft-lbs down low. Equipped with shorty headers, the Kenne Bell supercharged LS3 produced 632 ft-lbs of torque, but this jumped to 645 after the installation of the long-tube Hooker headers.

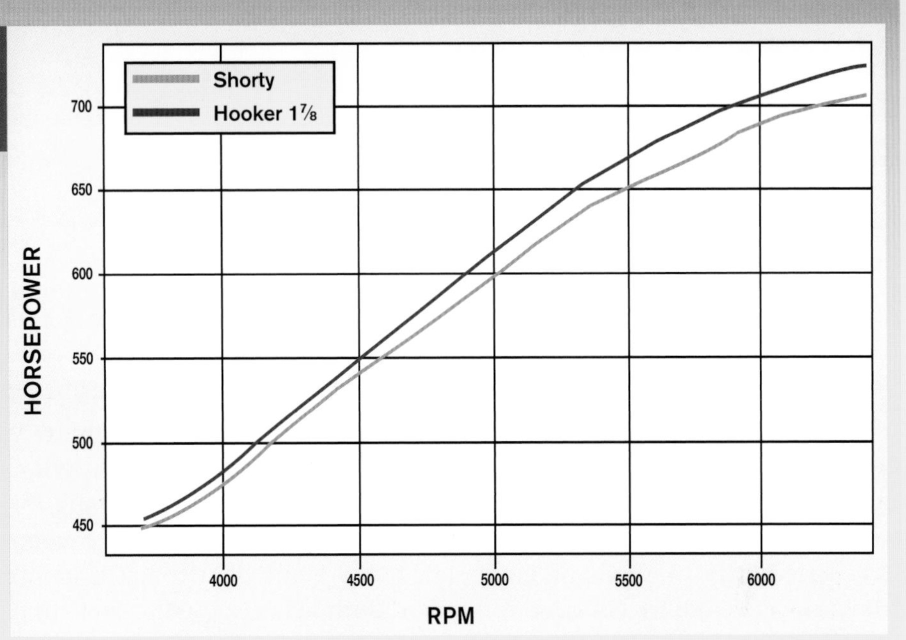

Shorty vs 1⅞ Hooker KB SC LS3 (Torque)

Short Extension: 632 ft-lbs @ 4,400 rpm
Hooker 1⅞ Headers: 645 ft-lbs @ 5,100 rpm
Largest Gain: 20 ft-lbs @ 5,700 rpm

Even more than I like seeing power gains, I like to see power gains through the entire rev range. Note that the long-tube headers offer scavenging improvements at lower engine speeds, even on the supercharged application. The power gains increased with engine speed, as the headers improved the power output of the supercharged LS3 by 21 hp.

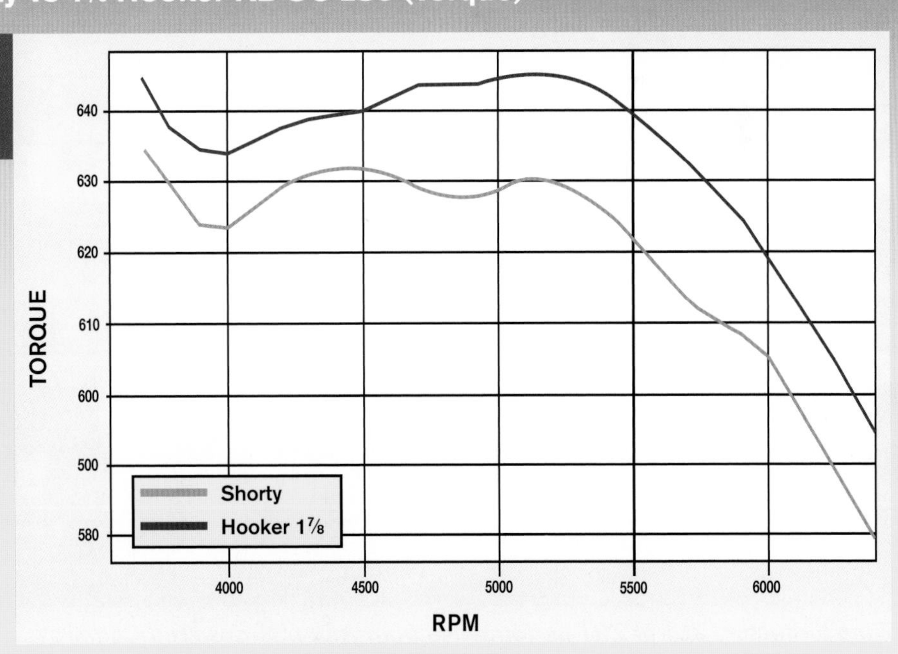

CHAPTER 5

SUPERCHARGING

When it comes to making serious power, nothing beats forced induction, which covers both superchargers and turbochargers. In this chapter, I look specifically at supercharging (turbos are covered in Chapter 6). Under the supercharger heading are basically three varieties: Roots, centrifugal, and twin-screw. They all provide different response rates and they also differ in their levels of efficiency.

Enthusiasts are often looking for the proverbial "best" supercharger, but know right off the bat that no such animal exists. The fact that many systems are currently offered for LS applications means that there is certainly at least one system that meets your needs. Forget bragging about how your system is better than everything else on the market and concentrate instead on how much fun it is behind the wheel (or even making the system better).

Superchargers are popular for a number of reasons, including boost response, packaging, and even emissions certification. Unlike turbos, which must rely on exhaust energy to spin their turbine wheel, superchargers are mechanically coupled to the engine. As such, superchargers (especially the positive displacement variety) have the unique ability to provide immediate boost response. This boost response offered by the Roots and twin-screw superchargers equates to massive low-speed torque production. Stick your foot in the throttle of a supercharged LS3 and you are instantly rewarded with acceleration (or possibly tire spin).

Although we all brag about peak torque and horsepower numbers, it is this immediate torque production that gets that hefty Camaro rolling. This instant gratification can be used on a daily basis (even at part throttle), without having to rev the engine to redline to find the power. It's no wonder that General Motors selected

Looking for factory ZL1 or ZR1 performance from your LS3-powered Camaro? This twin-screw Kenne Bell supercharger easily bettered the factory-supercharged applications, and the 2.8 was just getting started. Capable of supporting more than 1,000 hp, the blower just needed something more than a stock engine to apply boost to!

positive displacement superchargers to power the heavy-hitting CTS-V, ZL1, and ZR1 applications.

Superchargers for LS applications are available from a variety of sources. Whether OEM or aftermarket, systems can be divided into three designs: Roots, twin-screw, and centrifugal. The OEMs (CTS-V, ZL1, and ZR1) and those from Edelbrock and Magnuson fall into the Roots category. Twin-screw superchargers are available from both Kenne Bell and Whipple. The centrifugal camp includes Vortech, Paxton, and Procharger.

Both the Roots and the twin-screw are classified as positive displacement superchargers, meaning that they provide a fixed amount of airflow for each revolution. The displacement can be changed by altering the size of the casing and attending rotor pack to meet specific flow and power goals; the amount of boost can be adjusted by increasing rotor speed relative to engine speed. This is accomplished by changing the drive pulley ratio (blower pulley versus crank pulley size).

Boost and blower sizing are also adjustable on the centrifugal supercharger, but the flow rate from the supercharger increases with impeller speed. Centrifugal superchargers tend to be more effective at higher engine and blower speeds, but they lack the boost response of positive displacement superchargers.

In addition to boost response, centrifugal and positive displacement superchargers also differ in their intake manifold design. Because a centrifugal supercharger is mounted (much the same as other accessories) to the side of the engine (block and/or head), they discharge boost through the factory or aftermarket intake. By contrast, the central mounting position (between the heads) of the typical Roots or twin-screw supercharger requires an integrated intake (and usually intercooler).

Packaging the blower, intercooler, and intake under the tight confines of the factory Camaro or Corvette hood leaves little room for runner length; thus most positive displacement systems tend to run very short intake runners. As you saw in Chapter 1, decreased runner length has a negative effect on low-speed and mid-range torque production. Although the centrifugal cannot match the positive displacement for low-speed boost response, part of the deficit is overcome with the use of proper runner length. The tuning effect of runner length happens regardless of the presence of boost. A good rule is that if it makes more power naturally aspirated, it will make more power with the forced induction.

Both centrifugal and positive displacement systems employ intercooling to help lower the charge temperature. The unfortunate side effect of positive pressure (boost) is unwanted heat. Intercooling is used to help rid the inlet air of some of that unwanted heat. Since the inception of the supercharger, enthusiasts (more likely engineers) started looking for ways to cool the charge temperature on supercharged engines.

Heat should be considered the enemy of performance. Hotter air contains fewer oxygen molecules per volume, so that volume is able to support less horsepower. Provide an engine with cold, dense (oxygen-rich) air and watch the power increase.

Heat is also one of the variables that determine the detonation threshold of a given cylinder. Ignition timing, octane rating, and chamber design all fall into the detonation threshold variable category as well. Since heat is bad, and air is heated when compressed by superchargers, intercooling is used to both increase power and lower the likelihood of detonation.

Adding boost from a Vortech supercharger to an LS3 or LS7 combination can more than double the power output.

Test 1: Kenne Bell ZL1 Upgrade at 13 and 18 psi

The ZL1 Camaro has a lot going for it, but one of the major highlights has to be the amazing 6.2 LSA engine. Factory equipped with an intercooled supercharger, the ZL1 was just one small step down from the even more amazing LS9-powered ZR1 Corvette. Rated at 580 hp in the Camaro, the LSA engine had even more power potential. Although the factory 1.9 supercharger could be turned up for more power, eventually it became the limiting factor in power production.

The key to increased power on the LSA was to perform a supercharger upgrade from Kenne Bell. Kenne Bell offered a number of displacement superchargers, but this test involved the installation of a 2.8 twin-screw capable of supporting more than 1,000 hp. In addition to the powerful supercharger, the Kenne Bell kit also featured an intercooler and induction upgrade to ensure the blower received an unrestricted source of cold air.

This particular ZL1 was equipped with an automatic transmission, so the power numbers generated on the Dynojet were slightly lower than those of a manual transmission. The first step of the test was to run the LSA in stock trim to establish a baseline.

Run as delivered from Chevrolet, the supercharged 6.2 LSA produced 466 hp and 496 ft-lbs of torque (SAE). Using the factory flywheel rating, you see that the automatic drivetrain was soaking up roughly 20 percent of the available power. The factory Eaton supercharger produced a peak of 7.9 psi during the testing.

After installation of the Kenne Bell twin-screw supercharger with a 3.75-inch blower pulley, the Kenne Bell supercharged LSA produced 561 rear-wheel hp and 550 ft-lbs of torque. The Kenne Bell 2.8 produced a peak boost pressure of 13 psi. After replacing the 3.75-inch blower pulley with a smaller 3.0-inch, the boost pressure jumped to 18 psi, whereas the LSA produced 668 hp and 636 ft-lbs of torque.

The key to the success of the Kenne Bell blower upgrade was the twin-screw supercharger. More than just a bigger blower (2.8 versus 1.9), the more efficient twin-screw also offered reduced parasitic losses and lower charge temperatures.

The 580-hp 6.2 ZL1 is an impressive performer. If you see one carrying a Kenne Bell banner, you know that it has a lot more power than stock.

Kenne Bell ZL1 Upgrade at 13 and 18 psi (Horsepower)

Stock ZL1 Camaro (8 psi): 456 hp @ 6,300 rpm (8 psi)
KB ZL1 Camaro (13 psi): 560 hp @ 6,200 rpm
KB ZL1 Camaro (18 psi): 668 hp @ 6,300 rpm
Largest Gain: 216 hp @ 6,300 rpm

To illustrate the gains offered by the Kenne Bell twin-screw supercharger, I ran the automatic ZL1 first in stock trim. Plenty powerful, the supercharged 6.2 produced peak numbers of 456 hp and 496 ft-lbs of torque (SAE) at the wheels. Although capable of more, the stock Eaton supercharger produced a peak boost reading of 7.9 psi. Not surprisingly, installation of the Kenne Bell twin-screw supercharger increased the boost pressure. Running 13 psi of Kenne Bell boost, the LSA produced 560 rwhp and 550 ft-lbs of torque.

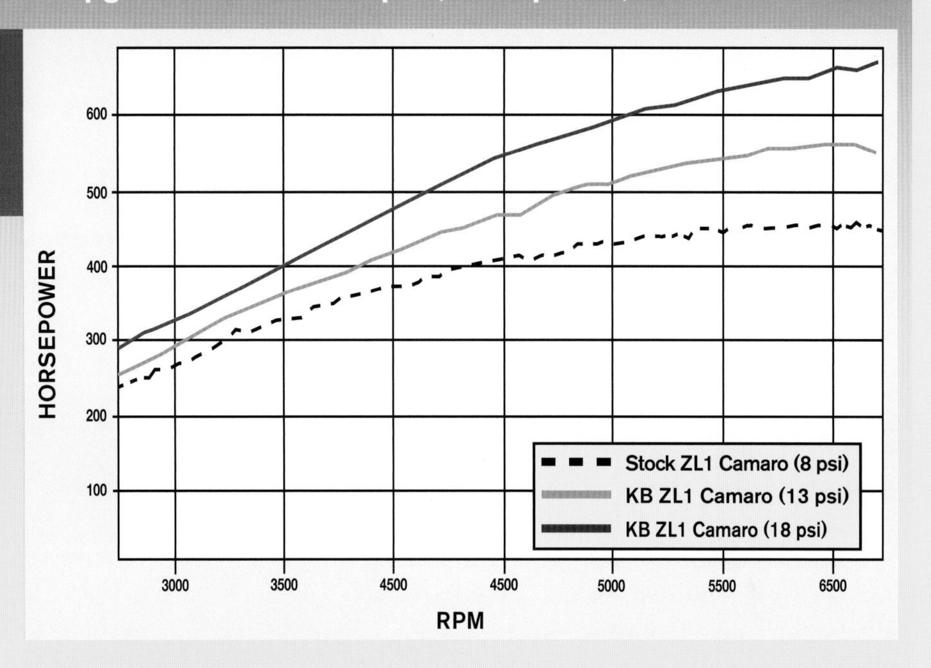

Kenne Bell ZL1 Upgrade at 13 and 18 psi (Torque)

Stock ZL1 Camaro (8 psi): 496 ft-lbs @ 3,800 rpm
KB ZL1 Camaro (13 psi): 550 ft-lbs @ 4,200 rpm
KB ZL1 Camaro (18 psi): 636 ft-lbs @ 4,500 rpm
Largest Gain: 176 ft-lbs @ 5,200 rpm

With boost so easy to come by, it was hard to resist cranking up the boost on the ZL1 Camaro. Having an efficient inter-cooler and race fuel made the choice a no-brainer. The extra boost came by swapping out the 3.75-inch blower pulley in favor of a 3.00-inch pulley. This increased the boost supplied by the twin-screw from 13 to 18 psi. The result of the increased boost pressure was a jump in power from a stock 496 ft-lbs of torque to 636 ft-lbs of torque at 18 psi. This is as high as I ran the boost on this otherwise stock LSA engine, but the Kenne Bell had plenty more power left in reserve.

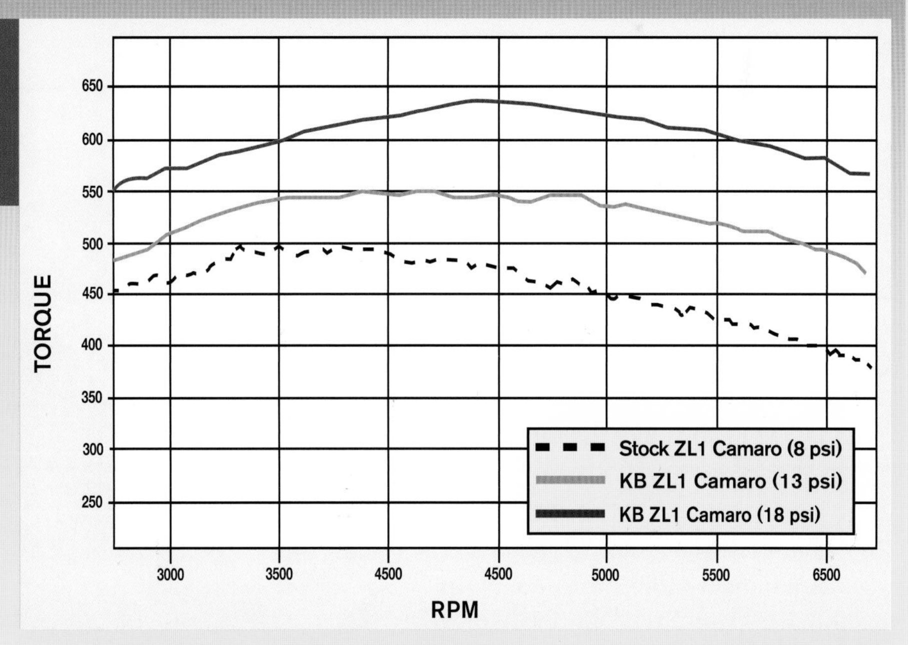

Test 2: Effect of Boost (Pulley Swap) on a Whipple Supercharged B15 LSX at 16 vs 23 psi

Boost from any supercharger is an amazing thing. The best thing about supercharging is the ability to easily increase the pressure to increase the power output. There is, of course, a limit to the amount of boost a supercharger can safely supply, but increasing the power output is as easy as changing the blower pulley.

Working with the crank pulley, the blower pulley helps determine the speed of the rotors (or impeller in the case of a centrifugal) relative to the engine speed. More rotor speed equals more boost pressure. Typically, more boost pressure equates to more power. Increasing the power output of an NA engine usually requires changing or adding a component such as cylinder heads, cams, or intake manifolds. Doing so on a blower engine, however, requires simply changing a pulley.

To illustrate the effect of a pulley change, I set up a B15 crate engine from Gandrud Chevrolet (GM Performance). The B15 designation indicated that the crate engine was able to withstand 15 psi of boost, but as is evident from this test, it can and has survived much more. The crate engine was upgraded with a BTR Stage IV blower cam (.617/.624 lift split, 239/258 duration split, and 119 LSA) and a Whipple 4.0 supercharger. The Supercharger was fed with a 105-mm (Ford) throttle body from Accufab and 120-pound Holley injectors. Holley also supplied the HP management system.

The Whipple was set up with a 4.75-inch blower pulley, and it produced a peak of 16.5 psi at 6,700 rpm. This produced a peak power reading of 880 hp and 715 ft-lbs of torque. After replacing the 4.75-inch blower pulley with a 4.00-inch version, the peak boost jumped to 23.3 psi where the blown B15 produced 991 hp and 827 ft-lbs of torque.

The test engine began life as a B15 LSX crate engine from Chevrolet Performance. Designed for boost, the GM crate engine featured a strong LSX iron block, forged internals, and six-bolt LSX heads.

The one change I made prior to testing was to replace the factory LSA cam with a Stage 4 blower grind from BTR.

Effect of Boost (Pulley Swap) on a Whipple Supercharged B15 LSX at 16 vs 23 psi (Horsepower)

Whipple Supercharged B15 LSX (4.75):
880 hp @ 6,700 rpm
Whipple Supercharged B15 LSX (4.00):
991 hp @ 6,700 rpm
Largest Gain: 120 hp @ 5,900 rpm

The great thing about any forced-induction application is the ability to easily increase the power output. Increasing the blower speed by changing the blower pulley resulted in a sizable jump in power through the entire curve. The pulley swap added as much as 7 psi of boost and increased the peak power output by 111 hp (from 880 to 991 hp).

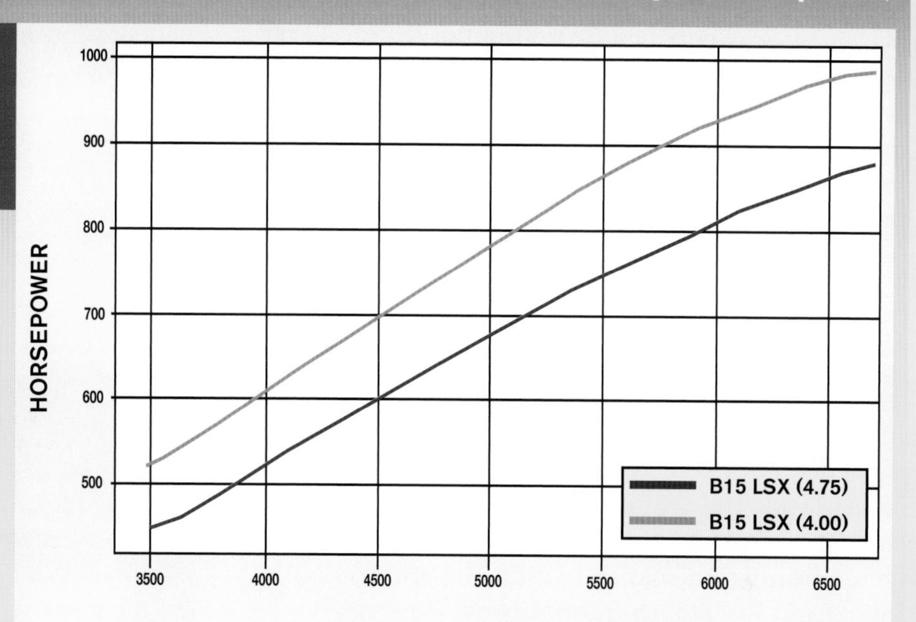

Effect of Boost (Pulley Swap) on a Whipple Supercharged B15 LSX at 16 vs 23 psi (Torque)

Whipple Supercharged B15 LSX (4.75):
715 ft-lbs @ 5,400 rpm
Whipple Supercharged B15 LSX (4.00):
827 ft-lbs @ 5,400 rpm
Largest Gain: 111 ft-lbs @ 5,100 rpm

Since the blower pulley swap increased the boost pressure through the entire rev range, the torque curve took a huge jump as well. We all like to talk about big peak power numbers, but the extra torque will really help accelerate the vehicle. The peak torque jumped from 715 to 827 ft-lbs, but the curve was up at least 100 ft-lbs everywhere.

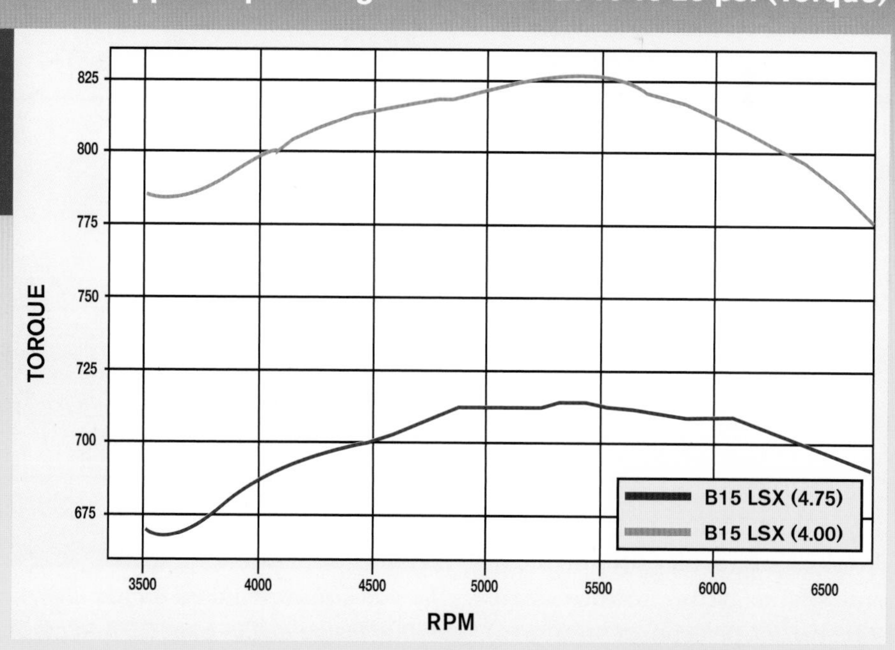

Test 3: 408 LS3 Hybrid Stroker: NA vs Vortech YSi at 13.5 psi

Not technically an LS3 or LS7, this test involved a 6.0 stroker equipped with LS3 heads. The 408-ci displacement was obtained by boring the 6.0 block .030 over to accept forged (10-cc dish) JE pistons and Total Seal rings. Extra displacement came from the 4340 forged-steel crank and matching connecting rods from Speedmaster. When combined with the 70-cc (CNC-ported) LS3 heads from Total Engine Airflow, the result was a static compression below 10.0:1.

The cam is critical on any performance build, so I chose a Comp Xtreme Energy grind that offered a .624-inch lift, 239/247-degree duration split, and wide 114-degree LSA. Comp Cams also supplied the matching hydraulic roller lifters, adjustable timing chain, and hardened pushrods. To keep all that wonderful boost sealed inside the cylinders, the TEA-ported heads were secured using ARP head studs and Fel Pro MLS head gaskets. Finishing touches included a 102-mm FAST LSXR intake and Big Mouth throttle body, along with a set of 1¾-inch headers and FAST XFI management system.

The heavy hitter of the test was obviously the Vortech centrifugal supercharger. This race kit featured a YSi supercharger capable of supporting more than 1,100 hp. The big blower was combined with a Mondo air-to-water intercooler and racing bypass valve. The bypass valve was necessary to eliminate the pressure buildup and compressor surge that occurs under hard deceleration. Given the intended power levels, I also installed a cog-drive system to eliminate belt slippage during testing.

To illustrate the gains offered by the YSi, I first ran the stroker LS3 hybrid in NA trim to establish a baseline. Before adding the Vortech, the LS3-headed 408 produced 591 hp and 554 ft-lbs of torque. Installation of the YSi and Mondo intercooler resulted in some serious power, to the tune of 1,005 hp at a peak boost pressure of just 13.5 psi. True to its design, the cog drive was slip-free, the intercooler kept the charge temperatures down (although I ran dyno water and not ice water), and the YSi was plenty happy at this four-digit power level (though the blower had another 200 hp left in reserve).

The big power producer was, of course, the Vortech YSi supercharger. Perfectly sized to exceed 1,000 hp on this stroker, the Vortech offered both flow and efficiency to help keep the power up and the inlet air temperatures down.

The 408 stroker featured forged internals from Speedmaster and JE pistons stuffed inside a .030-over 6.0 iron block. The stroker received a set of CNC-ported LS3 heads from TEA a Comp cam and FAST LSXR intake. Run in naturally aspirated (NA) trim, the stroker produced 591 hp and 554 ft-lbs of torque.

408 LS3 Hybrid Stroker: NA vs Vortech YSi at 13.5 psi (Horsepower)

NA 408 Stroker: 591 hp @ 6,300 rpm
Vortech YSi 408 Stroker (13.5 psi): 1,005 hp @ 6,600 rpm
Largest Gain: 420 hp @ 6,500 rpm

The 408 stroker was built with forged internals from Speedmaster and JE Pistons to withstand the power potential of the YSi. That the engine made respectable power in NA trim was part of the reason it was so easy to exceed 1,000 hp once I added the supercharger. The YSi was combined with a Mondo air-to-water intercooler to maximize power. Running a peak boost pressure of 13.5 psi, the Vortech YSi pushed the power output to 1,005 hp at 6,600 rpm.

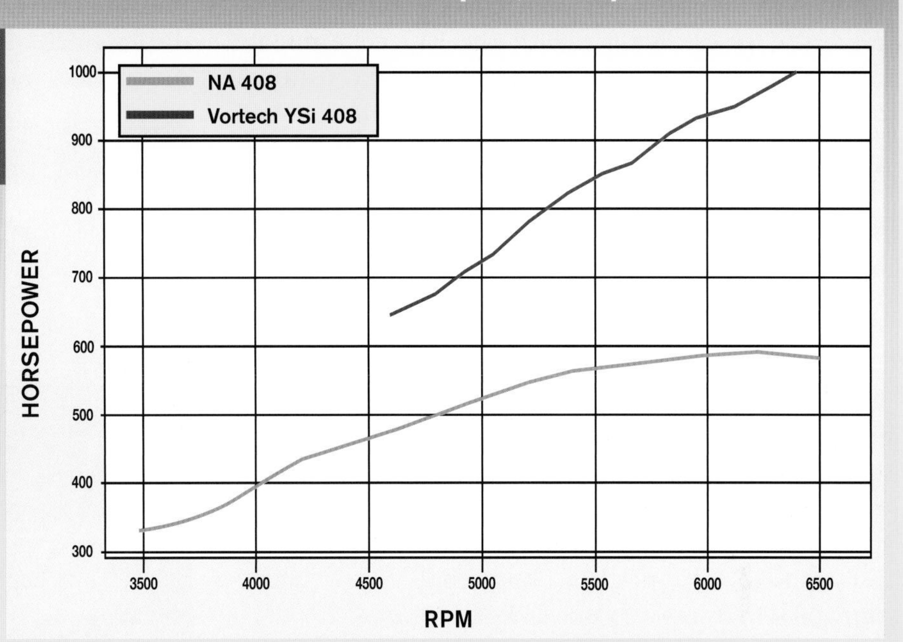

408 LS3 Hybrid Stroker: NA vs Vortech YSi at 13.5 psi (Torque)

NA 408 Stroker: 554 ft-lbs @ 5,200 rpm
Vortech YSi 408 Stroker (13.5 psi): 820 ft-lbs @ 5,900 rpm
Largest Gain: 394 ft-lbs @ 6,500 rpm

Unlike a positive displacement supercharger, the boost curve offered by the centrifugal Vortech YSi rose rapidly with engine speed. The supplied boost started at 6.6 psi at 4,600 rpm then rose to 13.5 psi at 6,700 rpm. Despite offering just 6.6 psi, the Vortech improved torque production by nearly 200 ft-lbs at 4,600 rpm and the torque gains increased with engine speed.

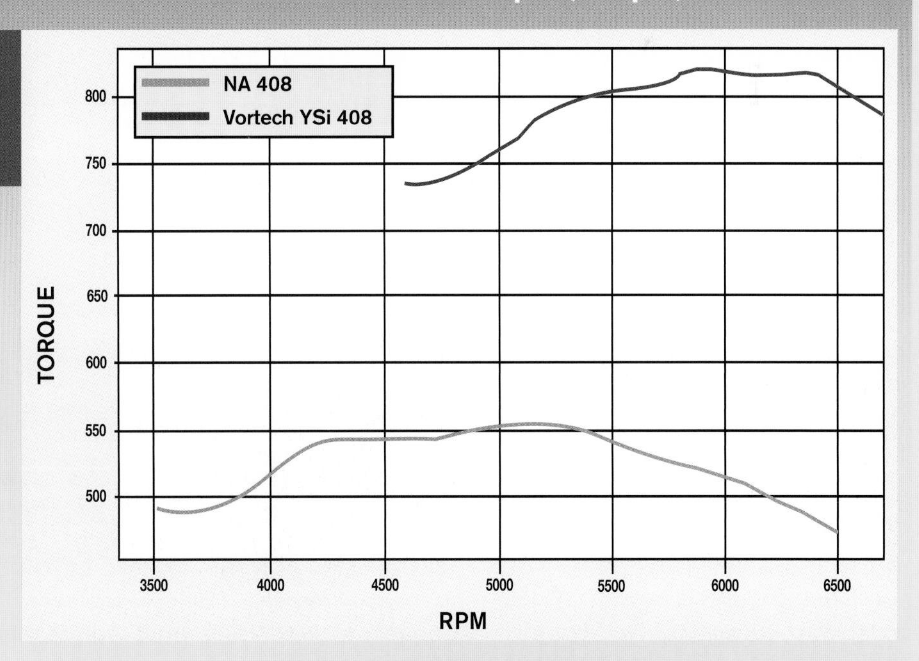

Test 4: LSX 376 B15-NA vs Magnuson TVS at 10.2 psi

On paper, both the GM B15 crate engine and Magnuson TVS hybrid-Roots supercharger offered some impressive specs. Starting with the B15 crate engine from Gandrud Chevrolet, the LS3-based 376 featured a forged-steel crank, powdered-metal rods, and forged aluminum pistons housed inside an iron LSX block. The crate engine was designed by General Motors for boosted applications up to 15 psi (the 15 in the B15 designation). In addition to the forged internals, the B15 was also blessed with six-bolt, LS3-based, LSX cylinder heads.

For this test, the engine was completed with a Milodon oiling system, FAST injectors, and a Holley HP management system. The LSX engine also featured a static (blower-friendly) compression ratio of 9.0:1 and a factory LS9 cam (.558/.552 lift split, 211/230 duration split at .050, and 122 LSA). Although it is not designed to run in NA trim, the B15 produced 479 hp at 6,000 rpm and 446 ft-lbs of torque at 5,100 rpm.

For this test, Magnuson supplied a Heartbeat hybrid-Roots supercharger. According to Magnuson, the kit (for a 2010 Camaro SS LS3) was designed as a 100-percent bolt-on, required no hood modifications, and offered a straight airflow path to maximize power and efficiency. In addition to the 2.3 supercharger, the Magnuson kit also featured an air-to-water intercooler to increase power while keeping detonation at bay. The hybrid rotor pack offered high thermal efficiency, 2.4:1 pressure ratio capability, and a bypass valve to eliminate parasitic loss under cruise conditions. These are important points because the dyno curves don't show every benefit offered by a particular supercharger.

With the B15 just begging for boost, I installed the Magnuson intercooled supercharger system and was immediately rewarded with 693 hp and 670 ft-lbs of torque. The system featured a drive ratio (blower and crank pulley) to produce a peak boost pressure of 10.2 psi. Consistent boost from the positive displacement supercharger ensured plenty of torque production and maximized the area under the curve.

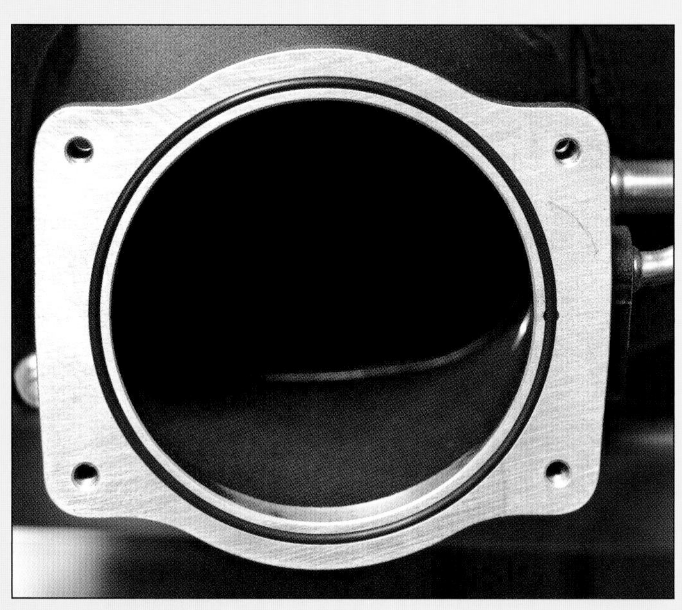

The front-mount throttle position eliminated the need to back-feed the supercharger.

Capable of supporting more than 900 hp on the right application, the Heartbeat supercharger was smaller, lighter, and more efficient than the already impressive TVS.

LSX 376 B15-NA vs Magnuson TVS at 10.2 psi (Horsepower)

NA 376 LSX B15: 479 hp @ 6,000 rpm
Magnuson 376 LSX B15: 693 hp @ 6,700 rpm
Largest Gain: 209 hp @ 6,400 rpm

As I have come to expect, the Magnuson Heartbeat supercharger offered immediate boost response and plenty of power on the B15 crate engine. Installation of the intercooled Magnuson kit increased the power output of the low-compression B15 from 479 hp to 693 hp at 10.2 psi.

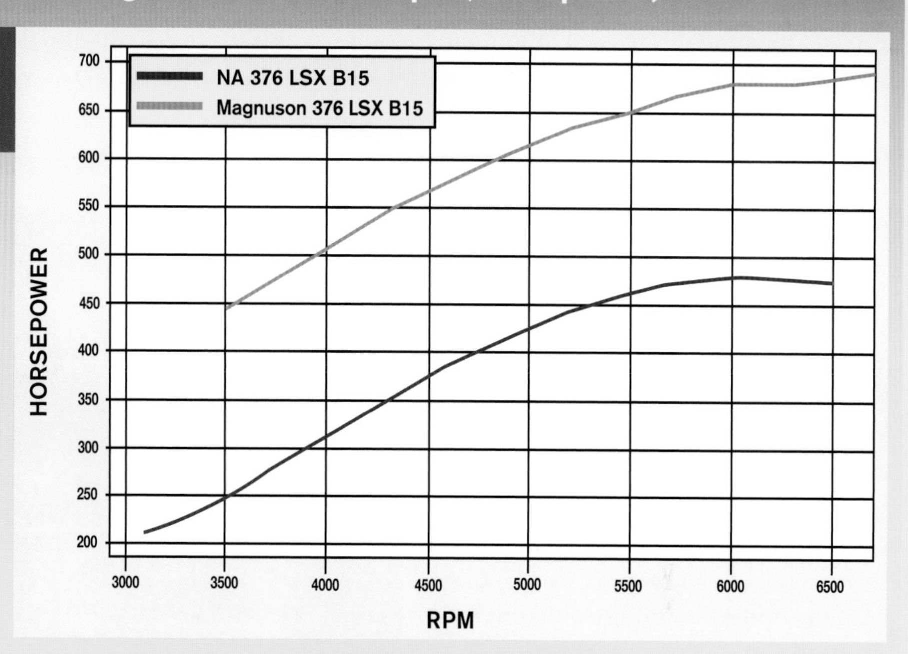

LSX 376 B15-NA vs Magnuson TVS at 10.2 psi (Torque)

NA 376 LSX B15: 446 ft-lbs @ 5,100 rpm
Magnuson 376 LSX B15: 670 ft-lbs @ 4,000 rpm
Largest Gain: 289 ft-lbs @ 3,600 rpm

The Magnuson Heartbeat test showed why positive displacement superchargers are so popular. The immediate boost response offered by the design provided sizable torque gains. The low-compression B15 was hardly a torque monster in stock trim, but offered big gains under boost. The Magnuson increased the torque output by almost 290 ft-lbs.

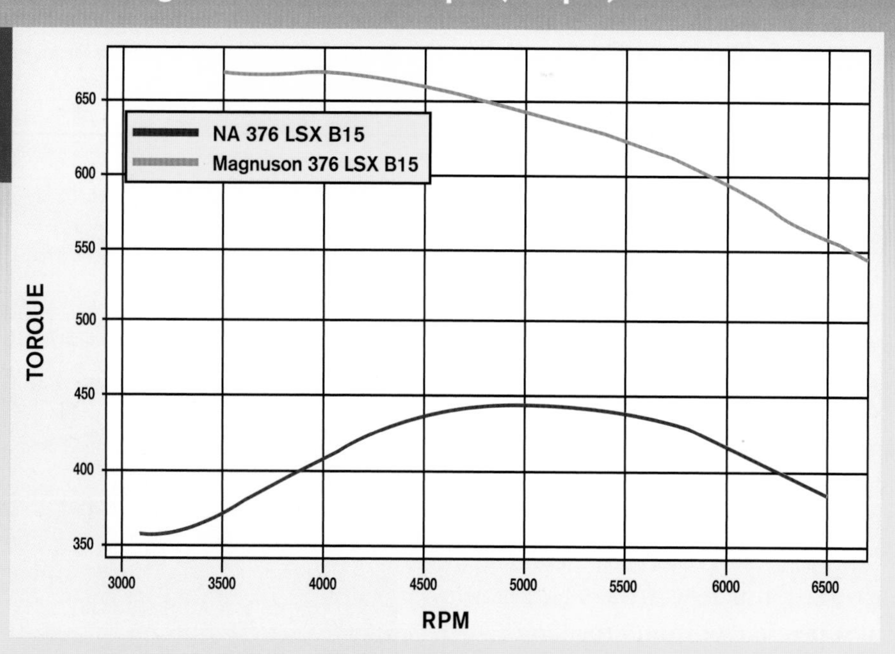

Test 5: NA 427 LSX vs Procharger F1A at 17 psi

I have stressed previously that there is no substitution for boost when it comes to producing extra power. Add a supercharger to just about any combination, even a stock one, and watch the power needle climb. The only thing better than a supercharged combination is a dedicated buildup, when every effort is made to enhance the power output of the normally aspirated engine before adding boost. Maximizing the power output of the normally aspirated engine allows the combination to produce any given power level with a lower boost level. An increase is power is always welcome, but doing so at a lower boost level means a reduction in inlet air temps-always beneficial.

Remember, with each extra pound of boost comes a commensurate increase in the inlet air temperature. Hotter air brings the combination ever closer to the detonation threshold, while simultaneously reducing power. Exceeding 1,000 hp at 20 psi is one thing, but doing so at just 15 psi is potentially much easier on the engine.

To illustrate what happens when you combine boost with a high-horsepower, normally aspirated combination, I added an F1A Procharger to no less than a 427 LSX. The 7.0L LSX featured forged internals from Lunati, CP, and Carrillo all stuffed neatly into a GM LSX block. The combo was topped with a set of TFS Gen X 260 LS7 heads and supplied a BTR Stage IV LS7 cam. Added to the package was an MSD Atomic intake, 105-mm Holley throttle body, and 80-pound Accel injectors. Of course the crowning touch was the F1A centrifugal supercharger from Procharger.

The kit included a highly efficient air-to-air intercooler and massive bypass valve. Once tuned, the normally aspirated 427 produced 654 hp and 593 ft-lbs of torque, but these numbers jumped significantly to 1,267 hp and 1,041 ft-lbs once I added 17 psi from the Procharger.

The heart of the Procharger supercharger kit was the F1A. According to Procharger, the efficient impeller design of the self-contained centrifugal supercharger was capable of supporting over 1,200 hp.

The 8-rib, serpentine belt used on the Procharger F1A kit drove the 3.85-inch blower pulley.

NA 427 LSX vs Procharger F1A at 17 psi (Horsepower)

NA 427 LSX: 654 hp @ 6,500 rpm
Procharger F1A 427 LSX (17 psi): 1,267 hp @ 6,400 rpm
Largest Gain: 614 hp @ 6,400 rpm

How else but with boost can you nearly double the power output of an already powerful 427 LSX? Adding the Procharger F1A to the 427 LSX dramatically increased the power output. Despite being rated near 1,200 hp by Procharger, the F1A-equipped 427 produced 1,267 hp at 17 psi of boost. This compared to just over 650 hp from the normally aspirated 7.0L.

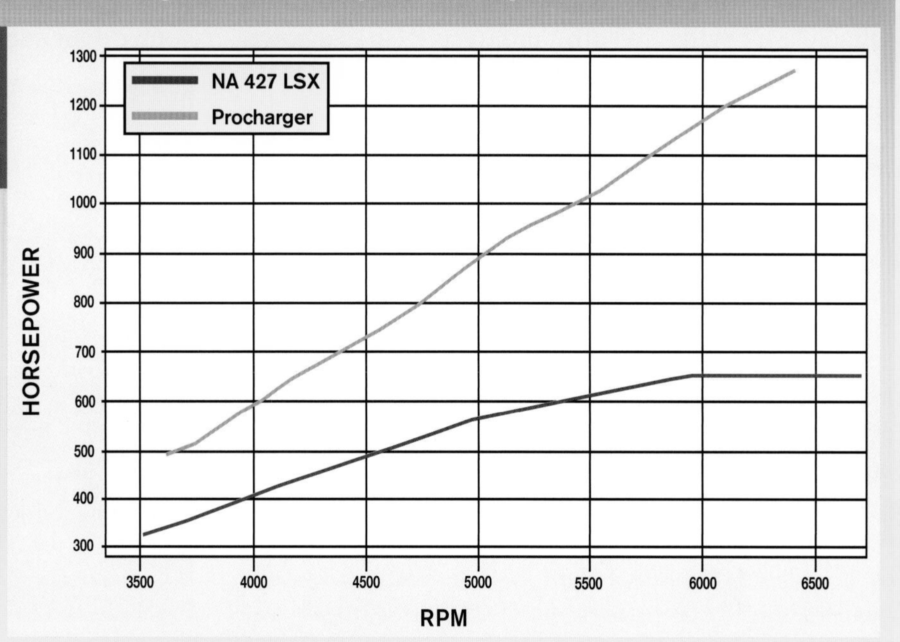

NA 427 LSX vs Procharger F1A at 17 psi (Torque)

NA 427 LSX: 593 lb-ft. @ 5,000 rpm
Procharger F1A 427 LSX (17psi): 1,041 ft-lbs @ 6,300 rpm
Largest Gain: 496 ft-lbs @ 6,300 rpm

The rising boost curve offered by the centrifugal supercharger produced a rising torque curve as well. Note that the normally aspirated 427 produced peak torque at 5,000 rpm, but the Procharger pushed peak torque to 6,300 rpm. What this means is that at the shut off point of 6,400 rpm, it was nowhere near the horsepower peak. With more than 1,000 ft-lbs of torque, this was on serious supercharged 7.0L.

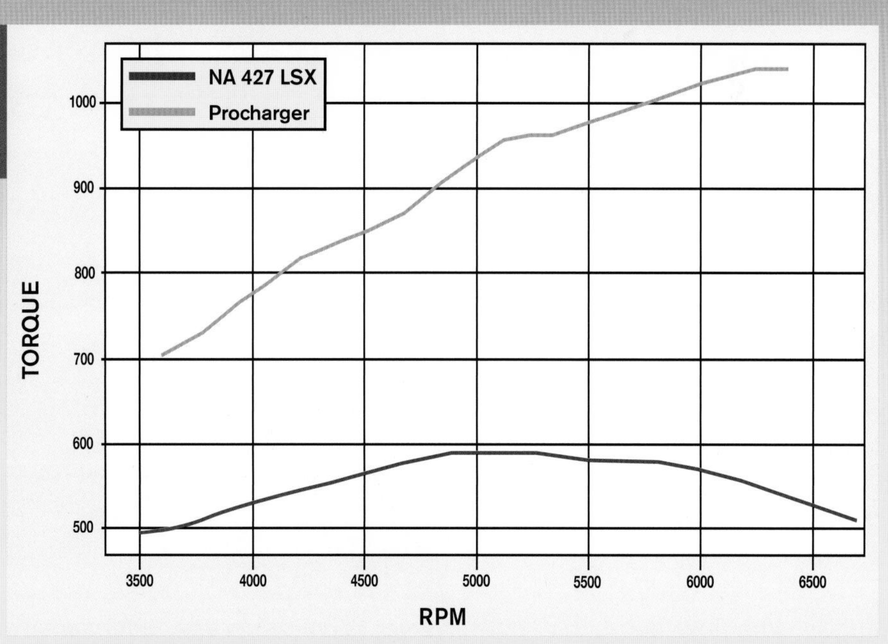

Test 6: 417 Stroker: NA vs Whipple 3.3 at 22 psi

Before I could run this high-boost blower test, I needed to resolve a few issues. The first was to secure a suitable test engine. To withstand the pressurized power offered by the Whipple supercharger, I assembled a 417 stroker based on an LS3 aluminum block. The LS3 stroker began as an LS3 crate engine from Gandrud Chevrolet. The 6.2 engine was upgraded with stroker components from Speedmaster, JE, and Total Seal. The 417 stroker was then equipped with a Comp cam (PN 281LR HR13) that offered a .617/.624 lift split, 231/239-degree duration split, and 113-degree LSA. Topping the stroker was a set of CNC-ported LS3 heads from Speedmaster. The big-chamber (81 cc) heads combined with the flat-top pistons to produce a blower-friendly compression ratio, but all testing was performed on Rocket Brand race fuel. Knowing that I was searching for four-digit power levels, I installed a set of Holley 150-pound injectors run by a Holley HP management system. Using long-tube headers, an ATI dampener, and FAST LSXR intake, the NA 417 stroker produced 576 hp at 6,100 rpm and 551 ft-lbs of torque at 5,100 rpm.

Before running the supercharger, I made a few adjustments to the as-delivered system. The inlet to the Whipple supercharger featured a dedicated throttle body flange with two different bolt patterns. Not surprisingly, the mounting flange accepted a standard GM throttle body, but I opted for something larger. Rather than run the smaller GM unit, I stepped up to a 105-mm Ford throttle body from AccuFab. Since high boost produces plenty of heat, I kept the charge temperatures down by employing the air-to-water intercooler. Because all that boosted air must exit the engine, I installed a set of 1⅞-inch headers from American Racing (run on the NA engine as well).

As always, I started out conservatively with low boost, but pulley swaps and tuning eventually resulted in a peak of 21.8 psi, where the supercharged stroker produced 1,047 hp and 984 ft-lbs of torque. Any time you exceed 1,000 hp, you know you've done something right.

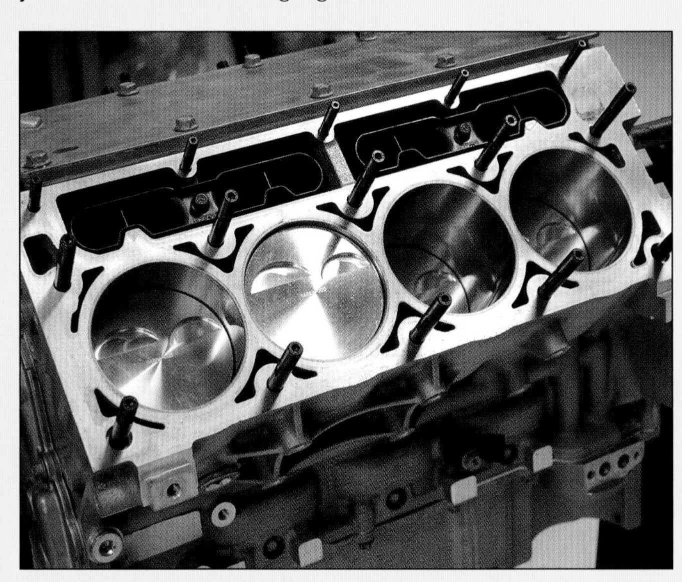

The LS3-based stroker featured a forged rotating assembly (SCAT, JE pistons, and Total Seal rings) stuffed inside an aluminum LS3 block supplied by Gandrud Chevrolet.

Tuning is critical on any performance engine, but especially on a high-boost stroker. I relied on a FAST XFI system to dial in the air/fuel and timing under boost. FAST also supplied the necessary injectors to feed this monster.

417 Stroker: NA vs Whipple 3.3 at 22 psi (Horsepower)

NA 417 LS3 Stroker: 576 hp @ 6,100 rpm
Whipple 417 Stroker (22 psi): 1,047 hp @ 6,500 rpm
Largest Gain: 476 hp @ 6,400 rpm

Each one of those graph lines represents 100 hp. The already-potent 417 LS3 stroker offered 576 hp, but things became serious once I added boost from the Whipple supercharger. Running 21.8 psi, the supercharger increased the power to 1,047 hp at 6,500 rpm.

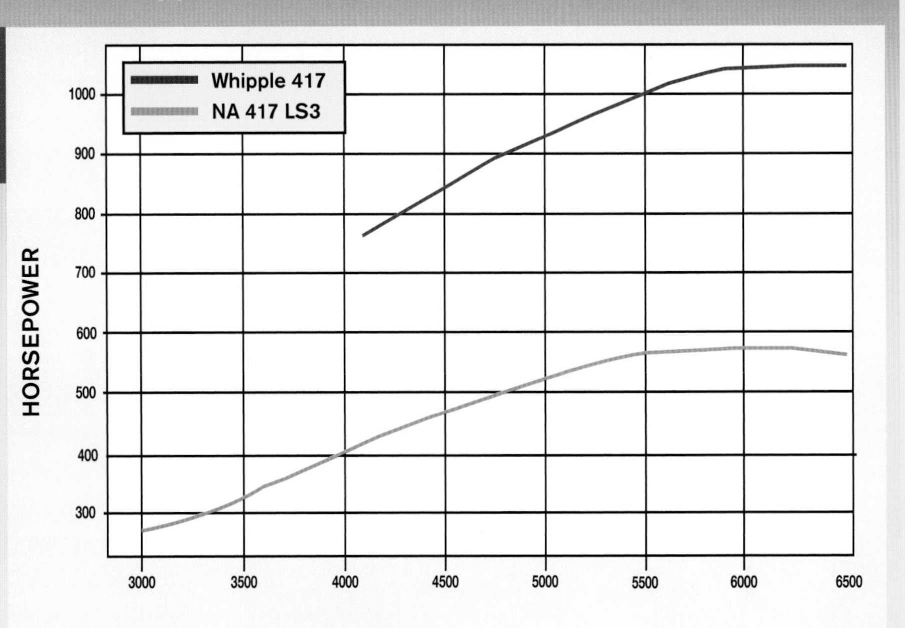

417 Stroker: NA vs Whipple 3.3 at 22 psi (Torque)

NA 417 LS3 Stroker: 551 ft-lbs @ 5,200 rpm
Whipple 417 Stroker (22 psi): 984 ft-lbs @ 4,800 rpm
Largest Gain: 433 ft-lbs 4,500 rpm

Torque is king for street cars and many other applications. How else can you add more than 400 ft-lbs of torque to a stroker? The Whipple supercharger offered a significant increase in torque through the entire rev range. With more than 550 ft-lbs, it was not as if the NA stroker was anemic, but with nearly 1,000 ft-lbs, the super stroker offered torque by the truckload.

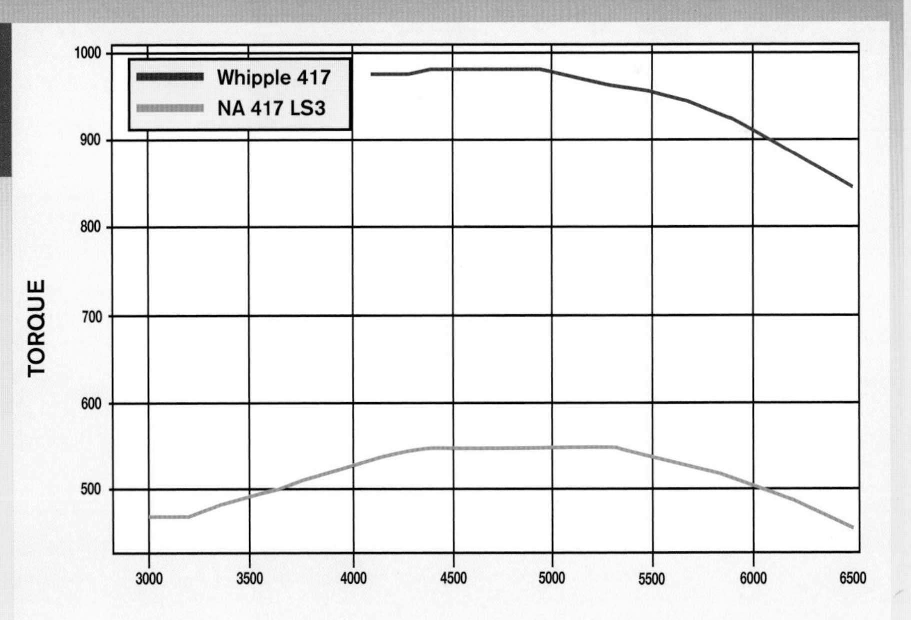

Test 7: LS3 and Stroker: NA vs Vortech at 7.7 and 13.3 psi

This test involved a pair of engines and two Vortech superchargers. The idea behind doubling up was to illustrate that power gains are available for almost any LS3 or LS7 combination, from stock to (in this case) stroker. The first engine that received the Vortech was a stock LS3 crate engine from Gandrud Chevrolet (GM Performance). The LS3 was equipped with long-tube headers and run without accessories using a Meziere electric water pump. I upgraded the stock injectors with a set of 75-pound injectors from FAST and relied on a Holley Dominator management system to program the air/fuel and timing values.

Run in stock trim, the NA LS3 crate engine produced 491 hp and 482 ft-lbs of torque. After installation of the Vortech T-Trim and Powercooler, the supercharged LS3 produced 710 hp and 610 ft-lbs of torque at a peak boost reading of just 7.7 psi.

Just how much power can you squeeze out of a Vortech supercharger?

The second test involved more of everything, including both cubic inches and boost. The LS3-based stroker combined a 4.065-inch bore with a 4.0-inch stroke. The SCAT crank and rods were combined with JE pistons and stuffed inside a new LS3 aluminum block. The stroker also featured a Comp cam that offered a .600-inch lift, 247/255-degree duration split, and 114-degree LSA. Topping the engine was a set of GM CNC L92 heads. The heads were upgraded with a set of double springs from BTR. This stroker relied on a FAST LSXR intake and throttle body as well as the same 75-pound FAST injectors used on the previous LS3.

This test engine was run with a FAST XFI management system. In NA trim, the stroker produced 585 hp and 541 ft-lbs of torque. After adding the Vortech YSi and Spearco air-to-water intercooler, power jumped to 949 hp and 807 ft-lbs of torque. The power was climbing rapidly at the shut-off point of 6,300 rpm, but not every test is run to maximize the power output. This was already about 100 hp more than the intended application required, so engine speed was limited to ensure a long, happy life (if only).

Run on the stock LS3 crate engine, the Vortech T-Trim offered a peak of 7.7 psi at 6,500 rpm. The result was an increase from 492 hp to 710 hp.

LS3 and Stroker: NA vs Vortech at 7.7 psi (Horsepower)

NA LS3 Crate Engine: 492 hp @ 6,100 rpm
Vortech LS3 Crate Engine: 710 hp @ 6,500 rpm
Largest Gain: 209 hp @ 6,300 rpm

What I like best about the power curve offered by a centrifugal supercharger is that it seems to want to continue to climb with engine speed. Despite a stock LS3 (with a mild stock cam), the power curve was still climbing at the shut-off point of 6,500 rpm. Credit the rising boost curve for the change where the engine makes peak power. With just 7.7 psi, the Vortech improved the power output of the LS3 by more than 200 hp.

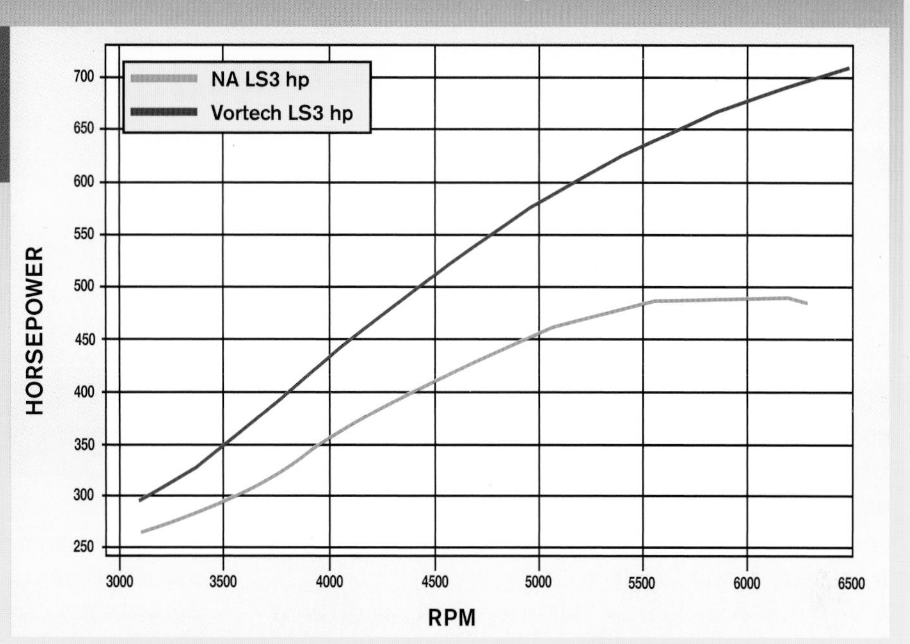

LS3 and Stroker: NA vs Vortech at 13.3 psi (Horsepower)

NA 413 Stroker: 585 hp @ 6,400 rpm
Vortech 413 Stroker: 949 hp @ 6,300 rpm
Largest Gain: 364 hp @ 6,200 rpm

There is no better combination than a bigger engine and more boost. Compared to the stock LS3, the stroker offered an additional 90 hp and the YSi was clearly more powerful than the T-Trim used on the crate engine. The combination of the two produced a power curve that was well on its way to eclipsing the 1,000-hp mark that had limited engine speed to just 6,300 rpm.

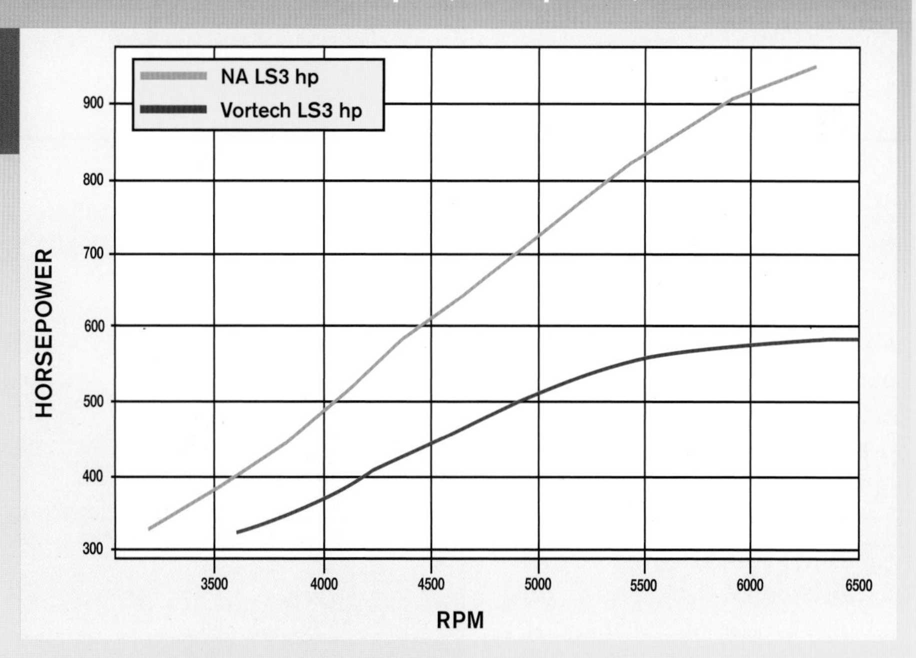

Test 8: 427 LS3 Stroker: NA vs Kenne Bell 3.6 at 21 and 26 psi

What happens when you combine a big engine with a big blower? The answer is obvious: You get big power. Why start with a big engine when you have a supercharger? It is true that boost from the supercharger greatly improves the power output of any engine, regardless of size, but in this case, bigger really is better. The great thing about increased displacement is that it becomes easier to produce power. Not only that, but the additional power comes at a lower relative engine speed than with a smaller engine.

The 427 stroker used for this test came from Turnkey and was the result of a 4.065-inch bore and 4.10-inch stroke. Unlike the more typical 4.125 x 4.0, this stroker offered more stroke than a stock LS7. The stroker featured a Comp cam (.624 lift, 235/243 duration split, and 113 LSA), GM CNC-ported LS3 heads, and a static compression of 8.5:1. Run in NA trim, the stroker produced 582 hp and 539 ft-lbs of torque.

For this test, Kenne Bell supplied one of its 3.6 superchargers. I know this was a serious piece of hardware, having witnessed it exceed 1,600 hp on a Ford modular engine (from Accufab). In addition to an efficient air-to-water intercooler, the 3.6 included a couple of interesting features. The twin-screw design featured a front seal to separate the gearbox from the rotor pack. The problem is that this front seal must resist the internal pressure supplied by the supercharger. Recognizing the potential of the situation, Kenne Bell designed the Seal Pressure Equalization (SPE) system, which effectively equalizes the pressure on both sides of the seal to eliminate the differential.

To this they added Liquid Cooling, which equalized the temperature differential between the inlet and outlet sides of the rotor pack to minimize distortion. These features, combined with the inherent efficiency of the (4x6) twin-screw design, helped ensure plenty of power potential. Applied to the 427 stroker, the result was 967 hp at 21 psi and 1,101 hp at 26 psi.

Not to be confused with intercooling (which the Kenne Bell kit featured), this Liquid Cooling referred to the system used to equalize the temperature differential between the cool (inlet) and heated (discharge) sides of the rotors.

This Kenne Bell 3.6 improved the power output of the 427 stroker by 521 hp.

427 LS3 Stroker NA vs Kenne Bell 3.6 at 21 and 26 psi (Horsepower)

NA 427 Stroker: 582 hp @ 6,300 rpm
KB 427 Stroker (21 psi): 967 hp @ 6,000 rpm
KB 427 Stroker (26 psi): 1,101 hp @ 6,300 rpm
Largest Gain: 521 hp @ 6,400 rpm

Only with forced induction do you get these kinds of power gains. The low-compression 427 stroker provided a perfect starting point for the supercharger. In NA trim, the stroker produced 582 hp. Adding 21 psi of boost from the intercooled Kenne Bell stepped things up to 967 hp, while another 5 psi pushed it to 1,110 hp.

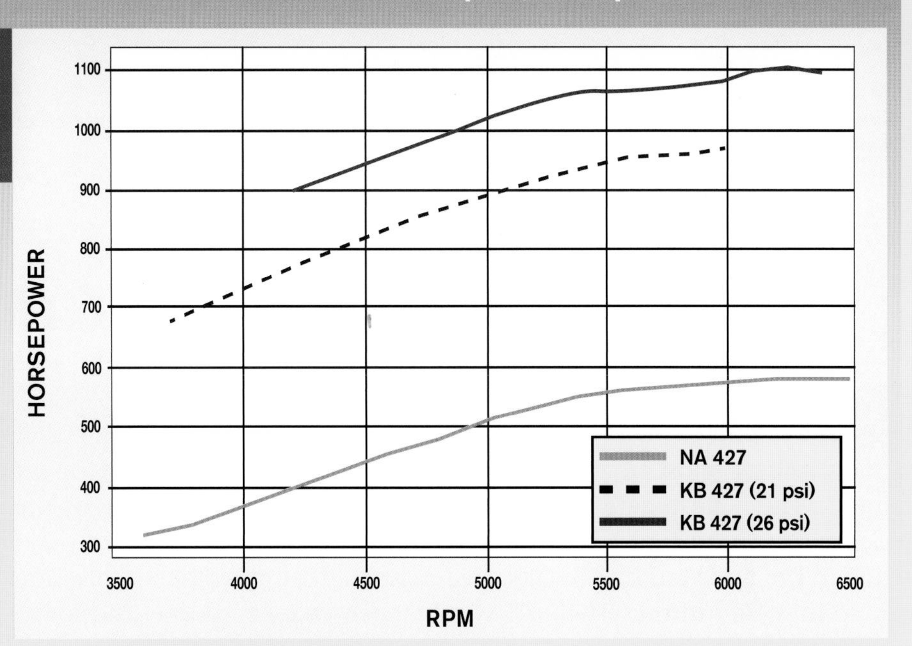

427 LS3 Stroker NA vs Kenne Bell 3.6 at 21 and 26 psi (Torque)

NA 427 Stroker: 539 ft-lbs @ 5,200 rpm
KB 427 Stroker (21 psi): 962 ft-lbs @ 3,700 rpm
KB 427 Stroker (26 psi): 1,124 ft-lbs @ 4,200 rpm
Largest Gain: 597 ft-lbs @ 4,300 rpm

Adding boost to the 427 LS3 stroker offered some serious torque gains. The Kenne Bell 3.6, twin-screw supercharger increased the torque output of the 427 stroker from 539 ft-lbs to 962 ft-lbs at 21 psi and up to 1,124 ft-lbs at 26 psi.

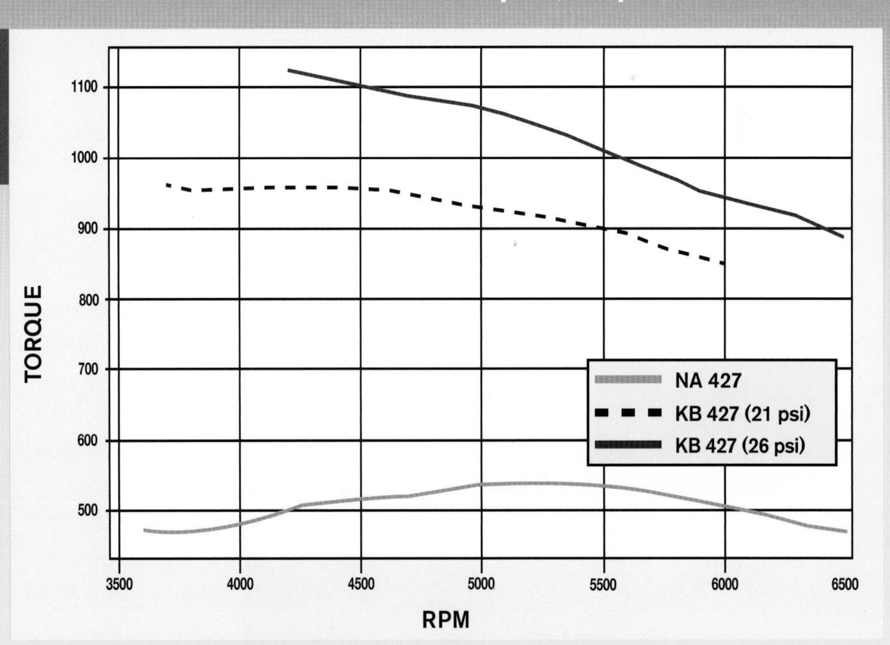

TURBOCHARGING

For maximum power production, it's hard to beat boost from a turbocharger. The cheap-date method for LS owners is to find a 5.3 truck engine in the junkyard then install a cam, springs, and turbo system. This combination has powered some serious street machines, and the same philosophy can be applied to LS3 and LS7 applications. Most serious LS3 (or LS7) efforts tend to be dedicated buildups. The question now is, How do turbo LS applications make such tremendous power?

Boost is really nothing more than a power multiplier. When I add boost from a turbo to a typical 430-hp LS3, it is important to understand that the NA LS is already running under boost, which comes courtesy of the atmosphere and equates to 14.5 psi at sea level. This atmospheric pressure obviously changes (as does the power output) with alterations in things such as elevation, temperature, and humidity, but the mechanics do not. As the piston races downward with the intake valve open, the external positive atmospheric pressure forces air into the negative pressure created by the piston.

This scenario creates plenty of power potential as you increase the external pressure applied to the engine above atmospheric. If an LS3 produces 430 hp at an atmospheric pressure of 14.5 psi, then you can theoretically double the power output (to 860 hp) if you supply an additional 14.5 psi of boost pressure. In truth, there are a number of reasons why this power/boost formula doesn't always work, but it is nonetheless a good indicator of potential power from a turbo engine.

When it comes to turbochargers, size really does matter. Like cam timing and intake manifold design, turbos should be selected to maximize power over a given RPM range, balancing response rate with ultimate boost and power potential.

Another great thing about the formula is that it can be applied at any given boost level. If you apply just 7.25 psi (.5 bar or 50-percent atmospheric pressure), you get a corresponding 50-percent increase in power (430 to 645 hp). The same goes for running 10 psi (430 x 1.689 = 727 hp), or even 2 bar (29 psi), where your 430-hp LS3 becomes a 1,290-hp monster. This example illustrates the importance of combining a powerful NA combination with boost because the power gains are simply multiplied by the original output. The more you start with, the more you finish with. Having more power to start with also allows you to reach any given power level at a lower boost level.

Although the boosted power output is a function of the original power multiplied by the boost (actually pressure ratio), know that all boost is not created equal. The advantage turbochargers have over superchargers is that very little power is required to drive the compressor of the turbo. The impeller or rotors of a supercharger are driven directly off the crankshaft. This mechanical coupling can provide immediate boost response, especially with positive displacement superchargers. However, as with the power steering, A/C, and alternator, the parasitic losses associated with driving the supercharger reduce the power output offered by the engine. This means that for nearly any given boost level, the turbo should produce more power than a comparable supercharger.

This power differential increases with boost (and flow), but know that 10 psi from a supercharger does not produce the same power curve or output as 10 psi from a turbo. The sacrifice for this efficiency can be boost response, but proper sizing can produce amazing results because factory turbo engines are able to provide peak boost pressure as low as 1,800 rpm (lower than you would want for almost any performance application).

Turbos have offered this type of performance since their inception, but one of the major reasons for a sudden surge in popularity is availability. Like it or not, the advent of affordable, offshore products has helped create the current turbo craze. Before the China connection, turbo kits were few and far between, primarily because of their expense. The average Joe could not or would not spend $5,000 to $6,000 on a turbo kit, but thanks to knock-off turbos, intercoolers, and the associated couplers and tubing, turbo pricing has dropped dramatically.

Obviously it pays to shop wisely, but putting together your own turbo kit can be done for less than half of what it cost not long ago and even less if you shop around. With $300 to $400 turbos, $125 intercoolers, and aluminum tubing bends readily available, it is possible to piece together a DIY turbo system for less than $1,000 if you start with factory exhaust manifolds. This type of kit is not going to put a scare in the Street Outlaw boys, but it is capable of boosting the power of an LS by 50 to 100 percent or more.

Another area where superchargers and turbochargers sometimes differ is in the intake manifold design. Turbos and centrifugal superchargers tend to use the factory (or equivalent) long-runner intake design. Positive displacement superchargers often replace the factory manifold to mount the blower. Primarily for packaging reasons, the supercharger is combined with some type of ultra short-runner intake because it is often difficult to get the supercharger, intercooler, and intake manifold under the hood of your average Camaro or Corvette.

There is some merit to the fact that the immediate boost response overcomes the torque losses (from charge filling) associated with optimized runner length, but using long-runner intakes is one advantage turbos (and centrifugal superchargers) have over positive displacement superchargers. As indicated in Chapter 1, runner length is one of the major factors that shape the entire power curve. Even on turbo applications, selecting the correct runner length tunes the combination to the desired engine speed. If you want your turbo LS3 or LS7 to run well up to 6,500 rpm, stick with a stock, FAST, or MSD Atomic intakes. If you are looking to elevate engine speeds, short-runner intakes such as the Holley Hi-Ram can push power production on a turbo engine past 7,000 rpm.

The heat generated by turbo systems needs to be managed properly. Make sure to shield components positioned near the exhaust system.

Whether running a supercharger or turbocharger, intercooling is an effective way to improve power and eliminate harmful detonation.

Because they control the boost pressure supplied to the engine, make sure to purchase quality waste gates such as this unit from Turbo Smart.

Test 1: Effect of Ignition Timing on a Turbo 4.8/LS3 Hybrid

Nothing wakes up an NA engine like a small dose of boost. The critical element when running boost is actually making sure the air/fuel and timing values are correct because a turbo engine runs thousands of trouble-free miles when treated to the proper tune. Typically, turbo engines require additional fuel and a slightly richer mixture than its NA counterpart. For maximum (safe) operation, an NA engine is typically tuned to achieve an air/fuel ratio near 13.0:1. By contrast, a turbo engine runs its best and is safer with a richer air/fuel ratio closer to 11.5:1. It is possible to run the turbo engine leaner than 11.5:1, but this is an effective air/fuel mixture for safe operation.

In terms of ignition timing, an NA engine runs best with more total timing than a forced-induction application. A good strategy is to have a drop of 1/2 to 1 degree of total timing per pound of boost. For the 6-psi application, this means a decrease of 3 to 6 degrees of total timing, but the actual amount is dependent on available octane.

To illustrate the power gains offered through changes in timing on a turbo LS, I installed a single Precision turbo on an LS3 equipped with a 4.8 crank. The turbo kit used a set of JBA headers feeding a custom Y-pipe equipped with a T4 turbo flange. The package also included a Turbo Smart waste gate, air-to-air intercooler, and 114-octane Rocket Brand race fuel. Set to run just 6 psi, the combination of the intercooler, safe air/fuel mixture, and race fuel allowed me to safely dial up the ignition timing.

Running 18 degrees of total timing, the turbo LS produced 539 hp and 502 ft-lbs of torque. Stepping up to 20 degrees resulted in 551 hp and 507 ft-lbs; 22 degrees brought 557 hp and 517 ft-lbs. The final test at 24 degrees resulted in 575 hp and 523 ft-lbs of torque. Each step up in timing brought additional power, but there is a limit to how far you can go with the available octane, and the timing increased peak torque less than peak power (less timing is required at the torque peak than the horsepower peak).

Precision supplied a 67-mm turbo for this low-boost test.

Optimizing ignition timing is important for an NA LS application, but it is critical on a turbo engine.

Effect of Ignition Timing on a Turbo 4.8/LS3 Hybrid (Horsepower)

18 Degrees: 539 hp @ 6,200 rpm
20 Degrees: 551 hp @ 6,200 rpm
22 Degrees: 557 hp @ 6,300 rpm
24 Degrees: 575 hp @ 6,200 rpm
Largest Gain: 36 hp @ 6,100 rpm

Increasing the total timing from 18 to 24 degrees increased the power output of the turbo LS from 539 to 575 hp. This shows the importance of ignition timing on a turbo application, but don't get greedy or you will just as quickly ruin a perfectly good engine.

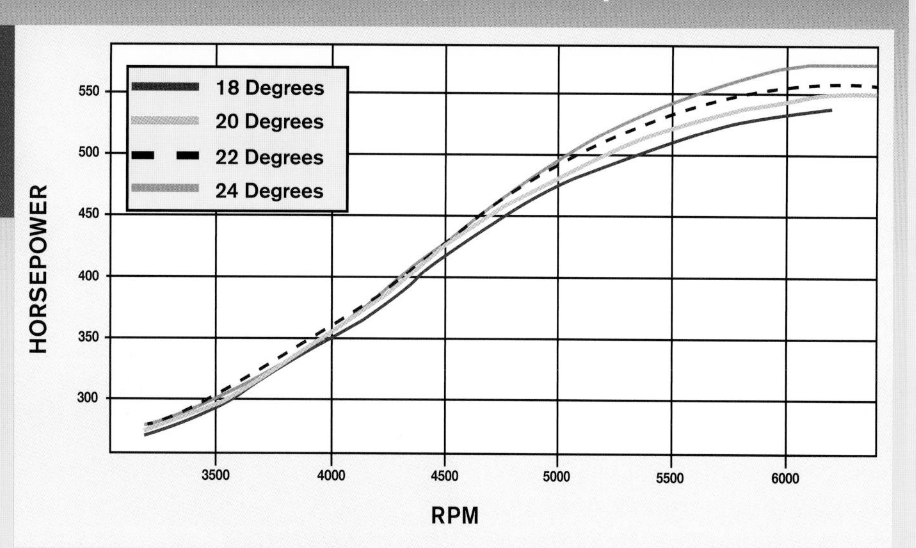

Effect of Ignition Timing on a Turbo 4.8/LS3 Hybrid (Torque)

18 Degrees: 502 ft-lbs @ 4,900 rpm
20 Degrees: 507 ft-lbs @ 4,900 rpm
22 Degrees: 517 ft-lbs @ 4,900 rpm
24 Degrees: 523 ft-lbs @ 5,100 rpm
Largest Gain: 26 ft-lbs @ 5,200 rpm

The change in ignition timing offered serious power gains on the turbo hybrid engine (actually all turbo engines). The additional ignition timing was more beneficial at higher engine speeds (typical for timing), but the torque gains were sizable as well. Care must be taken not to add too much timing because detonation can rear its ugly (and destructive) head. Running just 6 psi on race fuel and with an efficient intercooler allowed me to maximize timing on this turbo engine.

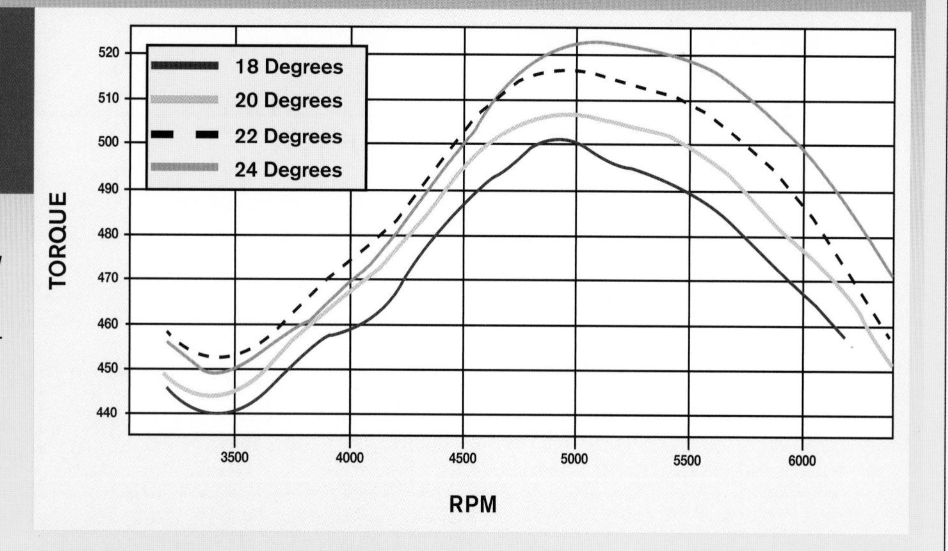

Test 2: 6.0 LS3 Hybrid: NA vs Single Turbo at 6.8 and 9.8 psi

This test involved turbocharging and applying boost to an LS hybrid. The hybrid was built by combining a 6.0 bottom end with an LS3 top end. The 6.0 short-block was boost-ready thanks to a forged rotating assembly that included a SCAT crank, K1 (6.125) rods, and JE Asymmetrical pistons. I combined a flat-top piston with the 70-cc combustion chambers for this turbo combination. The 6.0 hybrid was assembled using Fel Pro MLS head gaskets and ARP head studs. Although I ran a number of cams on this combination, this test was run with a stock LS2 cam. The stock LS3 heads were treated to a valvespring upgrade from BTR. I also ran 75-pound FAST injectors, the stock LS3 intake, and a FAST (manual) throttle body.

In essence, this 6.0 was an LQ9 or LS2 equipped with high-flow LS3 heads. Run on the dyno with a Holley HP Management system, long-tube headers, and a Meziere electric water pump, the LS produced 480 hp at 6,000 rpm and 472 ft-lbs of torque at 4,800 rpm. After adding the single turbo kit that included a Precision turbo (PN PT6766), CX Racing ATW intercooler, and Turbo Smart waste gate, I applied boost to the hybrid. Running 6.2 psi (6.8 psi at the torque peak), the turbo hybrid produced 615 hp and 619 ft-lbs of torque. After stepping up to 9.8 psi, the power output jumped to 718 hp and 687 ft-lbs of torque. The increase in boost improved power production through the entire rev range.

This hybrid started life as a 6.0 truck engine but was upgraded with JE pistons, K1 rods, and a SCAT crank. For this test I retained the stock LS2 cam.

The changes in ignition timing increased the power output of the turbo test engine by 36 hp.

6.0 LS3 Hybrid: NA vs Single Turbo at 6.8 and 9.8 psi (Horsepower)

NA: 480 hp @ 6,000 rpm
6.8 psi: 615 hp @ 5,500 rpm
9.8 psi: 718 hp @ 6,000 rpm
Largest Gain (9.8 psi): 238 hp @ 6,000 rpm

This test shows that even small amount of boost can have a dramatic effect on power production. Equipped with a 6.0 short-block, stock LS2 cam, and LS3 heads, the NA hybrid produced 480 hp at 6,000 rpm (as high as I revved it during the test). After adding the single Precision turbo, power jumped first to 615 hp at 6.2 psi (6.8 psi came at the torque peak) then to 718 hp at 9.8 psi. There was still more power to be had from the turbo, but this test wasn't designed to max out the engine or turbo.

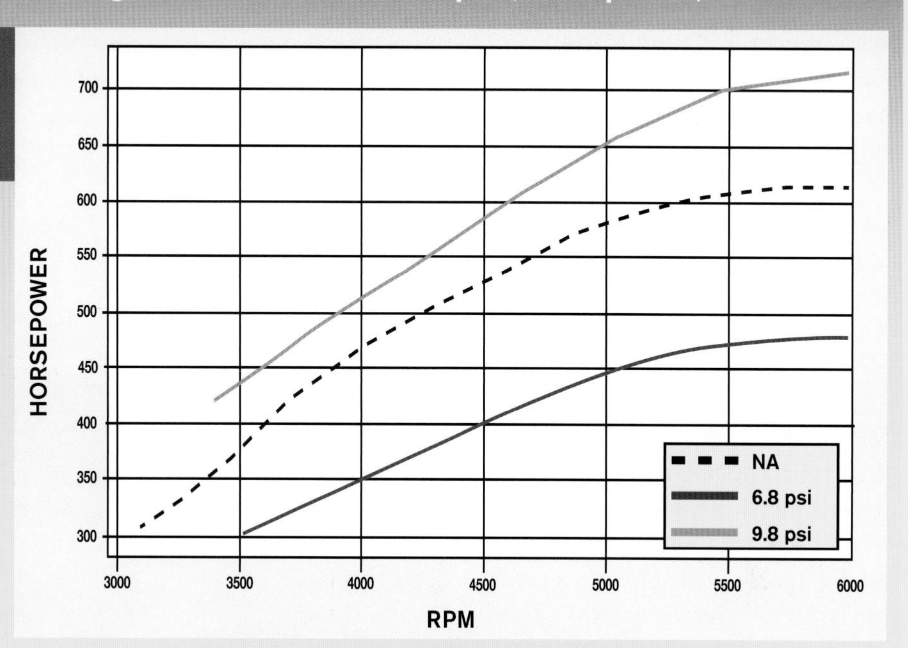

6.0 LS3 Hybrid: NA vs Single Turbo at 6.8 and 9.8 psi (Torque)

NA: 472 ft-lbs @ 4,700 rpm
6.8 psi: 619 ft-lbs @ 4,800 rpm
9.8 psi: 687 ft-lbs @ 4,800 rpm
Largest Gain (9.8 psi): 216 ft-lbs @ 4,800 rpm

The torque gains offered by the single turbo were impressive. Although the engine dyno artificially loaded the engine to provide a better boost curve than you might see on the street or strip, adding 6.8 psi (actually just 6.2 psi at the peak) of boost increased torque production from 472 ft-lbs to 619 ft-lbs. Adding another 3 psi pushed the torque peak to 687 ft-lbs.

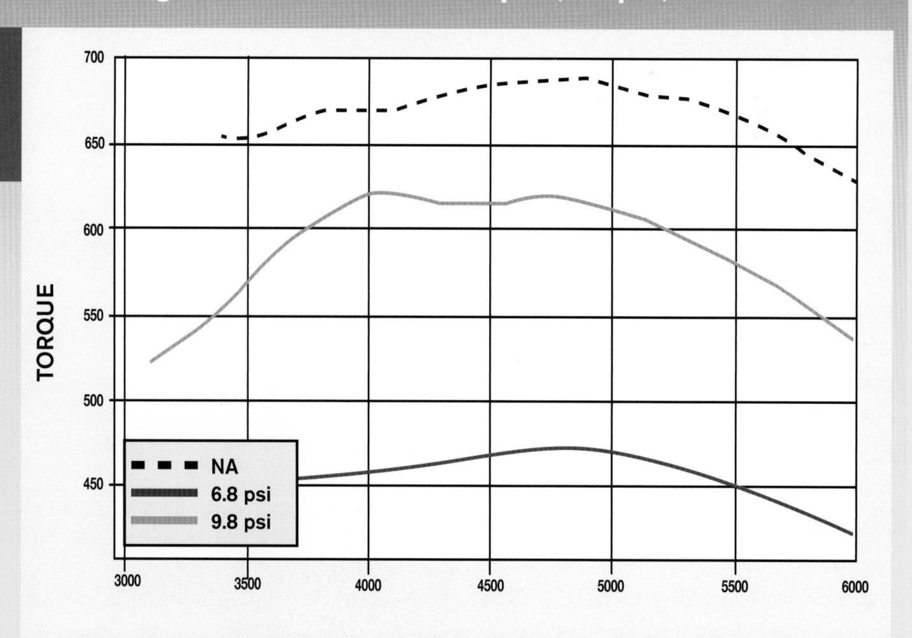

Test 3: Turbo Cam: LS9 vs BTR Stage II 4.8/LS3 Hybrid

The right cam is critical for any LS turbo application, including the short-stroke hybrid used for this test. Typical LS3 applications combine a 4.065-inch bore with a 3.622-inch stroke. This 6.2 combination works well and accommodates turbocharging. The combination used for this test replaced the stock 3.622-inch stroke with a smaller 3.267-inch stroke from a 4.8. With the exception of the smallest (4.8) and largest (7.0) engines in the family, all other LS engines (5.3, 5.7, 6.0, and 6.2) share the same stroke crank. The 4.8 shared the block with the 5.3, but the reduced displacement came from a shorter stroke. For this test, I combined the 4.8 crank with custom Lunati rods and forged JE pistons then stuffed it all inside an LS3 aluminum block. I then topped it off with a set of TS Gen X 255 heads and Holley Hi-Ram intake.

The turbo system consisted of a single 76-mm turbo from Precision Turbo fed by a pair of DNA turbo manifolds into a custom Y-pipe. Controlling the boost was a pair of Turbo Smart waste gates. Boost was fed through an air-to-water intercooler from CX Racing. This test was a comparison between the most powerful factory cam (LS9) and a Stage II turbo cam from BTR. The boost was limited to a maximum of 9 psi using just the waste-gate springs (no controller).

Run with an LS9 cam, the short-stroke turbo engine produced 701 hp and 598 ft-lbs of torque. The boost curve (on the spring) started at 8.5 psi, rose to a maximum of 9.4 psi, then dropped to 8.0 psi. After swapping in the BTR Stage II cam, the power output jumped to 733 hp and 621 ft-lbs of torque. The boost curve started at 8.2 psi, rose to 8.9 psi, then dropped to 8.0 psi. The BTR turbo cam improved peak power and offered more than 60 ft-lbs lower in the rev range.

Feeding the short-stroke turbo engine was a Holley High-Ram intake and TFS Gen X 255 heads.

What looks like an LS3 with forged pistons was actually a 4.8/LS3 hybrid. The aluminum LS3 block was stuffed with a 4.8 crank, custom Lunati rods, and JE forged pistons to produce a (high-RPM) short-stroke LS3.

Turbo Cam: LS9 vs BTR Stage II 4.8/LS3 Hybrid (Horsepower)

LS9 Cam: 701 hp @ 6,500 rpm
BTR Stage II Cam: 733 hp. @ 6,600 rpm
Largest Gain: 34 hp @ 6,600 rpm

The great thing about the BTR turbo cam was not just that it added 34 hp, but that it improved the power output through the entire rev range. More peak power is good, but more power everywhere is even better!

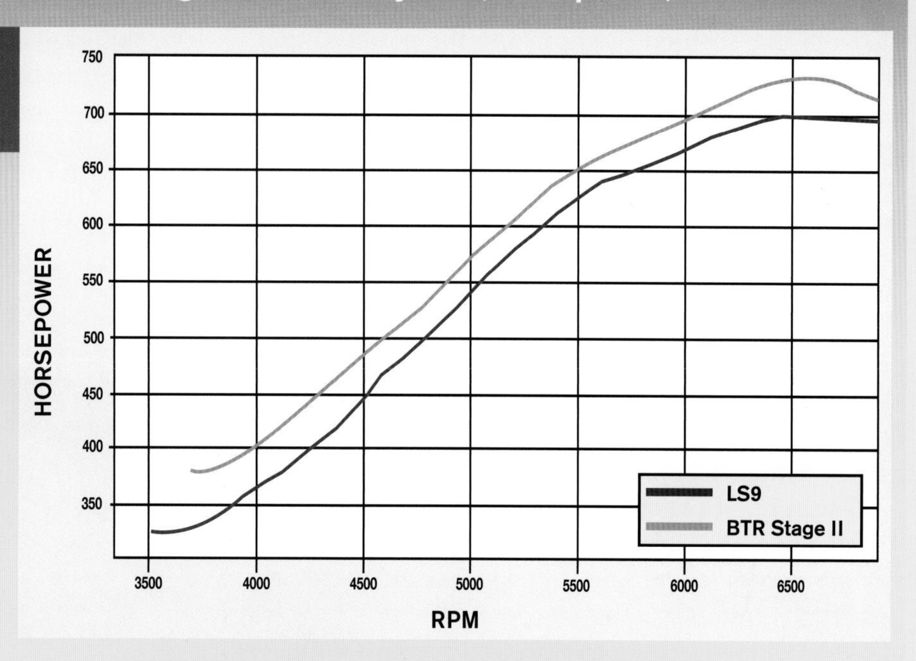

Turbo Cam: LS9 vs BTR Stage II 4.8/LS3 Hybrid (Torque)

LS9 Cam: 598 ft-lbs @ 5,600 rpm
BTR Stage II Cam: 621 ft-lbs @ 5,500 rpm
Largest Gain: 62 ft-lbs @ 3,800 rpm

We all love big horsepower gains but torque is more meaningful in real-world street driving. In addition to adding 34 hp, the Stage II BTR cam dramatically improved torque production over the LS9 cam. Down low, the BTR cam offered as much as 62 ft-lbs of torque, which is a sure indication that the BTR guys understand the needs of a turbo LS engine.

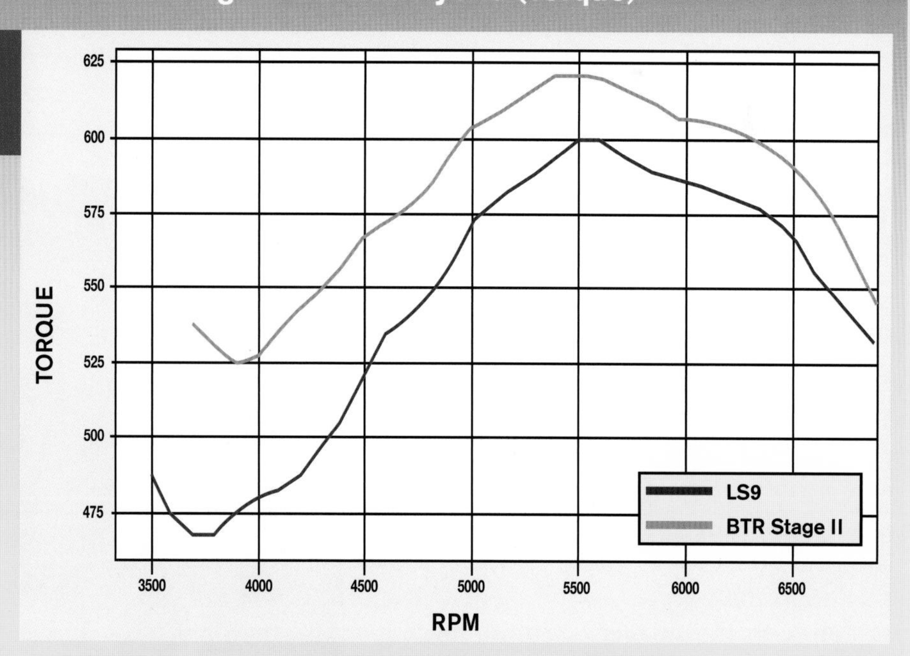

Test 4: Turbo Sizing: Big vs Small 76-mm

Turbocharging has been around since the birth of the internal combustion engine because it is a tried-and-true method of improving the power output. The basic concept is to force-feed the engine more air than it would ingest of its own accord. This air contains power-producing oxygen molecules, which when combined with fuel, ignite to provide the downward force on the crank. More air equals more oxygen, which in turn equates to engine power. Of course, this assumes the turbo is sized correctly for the intended application and desired power level. This also assumes you have sufficient fuel flow to meet the needs of the engine, something I neglected to do on this test (the results nonetheless demonstrate proper turbo sizing).

The 417 stroker test engine featured forged internals from Speedmaster and JE, along with an aluminum LS3 block from Gandrud Chevrolet. Topping the stroker was a FAST LSXR intake and throttle body, but I could only get my hands on a set of 60-pound injectors in time for testing. This ultimately limited the maximum power output, but the effect of turbo sizing was still clearly evident.

The 417 stroker was configured with a single turbo kit that consisted of a pair of tubular headers feeding a common Y-pipe. The Y-pipe was equipped with a pair of 45-mm Turbo Smart waste gates to control boost. The Y-pipe also featured a T4 turbo flange to readily accept a T4 turbo.

For this test, I ran a pair of T4 76-mm turbos, one from CX Racing and one from Precision Turbo. Although both advertised at 76 mm, the Precision unit was capable of supporting as much as 1,200 hp, and the CX turbo topped out under 800 hp. It bears mentioning that there was a substantial price difference between the two turbos ($450 to $1,800).

Run with the smaller CX turbo feeding an air-to-water intercooler, the turbo managed to produce just 7.3 psi at the power peak of 761 hp. The boost pressure rose as high as 9.9 psi early on, but fell off rapidly as it ran out of flow. The larger Precision turbo suffered no problem, but available fuel flow limited boost to just 12.5 psi, whereas the turbo stroker produced 913 hp and 925 ft-lbs of torque. The CX turbo would be great for a lower power level on a stock or mildly modified engine, but if you plan to crank up the boost on a stroker, better get the good stuff.

The test engine was a 417 stroker that included a Speedmaster 4.0-inch stroker crank and rods with a set of JE forged pistons. The stroker assembly was stuffed inside a new aluminum LS3 block from Gandrud Chevrolet.

A pair of ported LS3 heads from TEA feed the LS3 stroker. The heads were combined with a custom cam from BTR.

Turbo Sizing: Big vs Small 76-mm (Horsepower)

NA 417 Stroker: 576 hp @ 6,100 rpm
Small Turbo (7.3 psi): 766 hp @ 5,400 rpm
Big Turbo (12.5 psi): 913 hp @ 5,500 rpm (Fuel limited)
Largest Gain: 345 hp @ 5,600 rpm

Despite the lack of fuel flow, the graph shows gains offered by proper turbo sizing. Both are designated 76-mm turbos, but the Precision turbo offered significantly more flow than the unit from CX Racing. The boost pressure and power curve fell off rapidly at the top of the rev range with the smaller turbo.

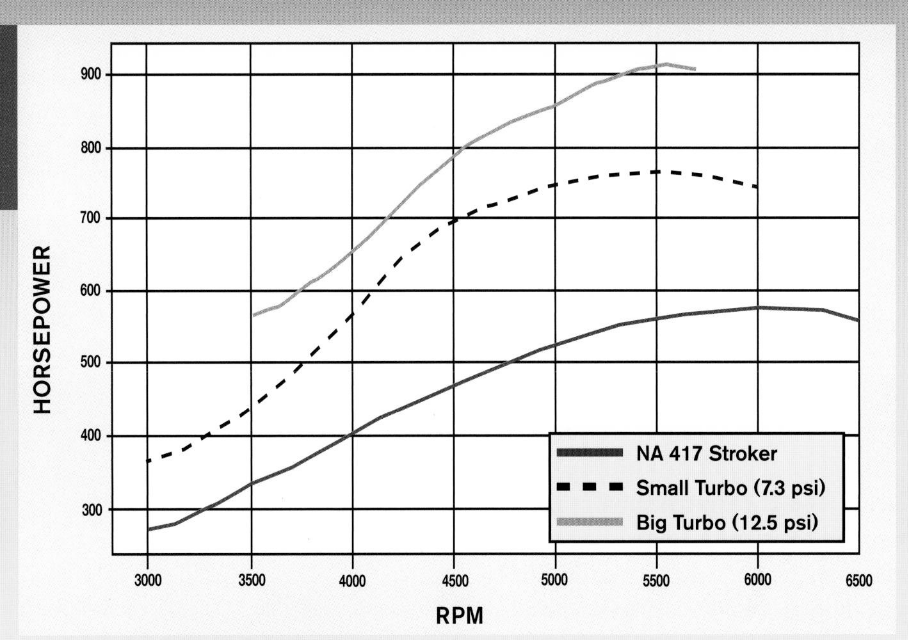

Turbo Sizing: Big vs Small 76-mm (Torque)

NA 417 Stroker: 551 ft-lbs @ 5,100 rpm
Small Turbo (7.3 psi): 815 ft-lbs @ 4,500 rpm
Big Turbo (12.5 psi): 925 ft-lbs @ 4,600 rpm (Fuel limited)
Largest Gain: 378 ft-lbs @ 4,600 rpm

The falling boost curve is even more apparent in the torque curves because the fall off in torque is even more pronounced. The smaller CX Racing turbo was about maxed out at 761 hp, but it managed to increase torque production by 260 ft-lbs over the NA engine. The Precision turbo was up more than 100 ft-lbs and only fuel flow kept me from easily eclipsing the 1,000 ft-lbs mark.

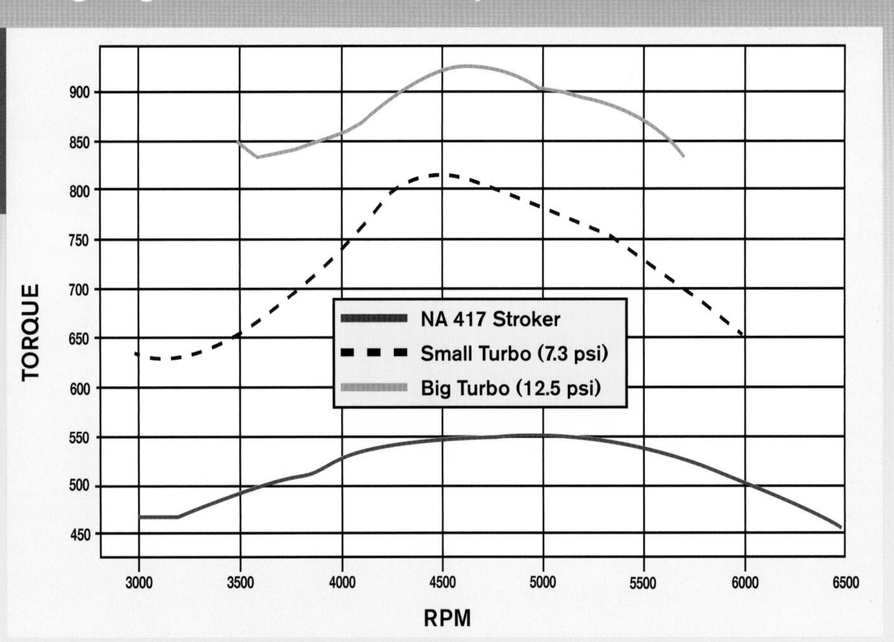

Test 5: Effect of Boost on a Turbo LSX B15 (14.6 vs 19.5 psi)

Boost from a turbo can be both a blessing and a curse. The blessing comes in the form of additional power because each extra pound of boost brings with it a substantial jump in power. The curse comes from the extra power as well because owners often become greedy after sampling all that wonderful power. If some boost is good, then more must be even better, right? Well, there is a limit to just how much fun is available without something getting hurt. The important point here is, don't get greedy.

This test was run on a GM B15, boost-ready crate engine from Gandrud Chevrolet. Equipped with forged internals and LSX LS3 heads, the crate engine was perfect for this boost test. Because the B15 came with an intake manifold, I installed a Holley Hi-Ram intake and FAST 102-mm throttle body, along with Holley 120-pound injectors. Tuning for the turbo combo was controlled by a Holley Dominator EFI system

The key to a successful turbo engine is knowing its limits. The tune is important, especially at elevated boost levels. Forged internals are slightly more forgiving in terms of detonation, but even the toughest components snap without the proper air/fuel and timing curves. I made sure to dial in the timing and air/fuel (kept constant for each boost level) using the Holley system.

Running the single Precision 76-mm turbo through the air-to-water intercooler, the turbo B15 produced 951 hp at 6,300 rpm and 891 ft-lbs of torque at 5,200 rpm. The boost curve started at 16.0 psi, rose to 17.5 psi, then fell to 14.5 psi at the power peak. I relied on a manual boost controller, but an electronic version would have kept the boost consistent. After cranking up the boost to 19.6 psi (at the power peak), the peak power numbers jumped to 1,083 hp and 979 ft-lbs of torque. Once again, the boost curve started out at 20.1 psi, rose to 22.7 psi, then dropped to 18.8 psi (19.6 psi at the power peak of 6,300 rpm).

Keeping things cool during testing was this single-pass, air-to-water intercooler from CX Racing. I ran dyno water through the core during testing.

Once again I relied on the single 1,200-hp 76-mm turbo from Precision.

Effect of Boost on a Turbo LSX B15 (14.6 vs 19.5 psi) (Horsepower)

Turbo LSX B15 (14.6 psi): 951 hp @ 6,300 rpm
Turbo LSX B15 (19.5 psi): 1,083 hp @ 6,300 rpm
Largest Gain: 132 hp @ 6,300 rpm

Boost always has a positive effect on the power curve, and this test was no different. Cranking up the boost from 14.6 to 19.5 psi increased the peak power numbers from 951 to 1,083 hp.

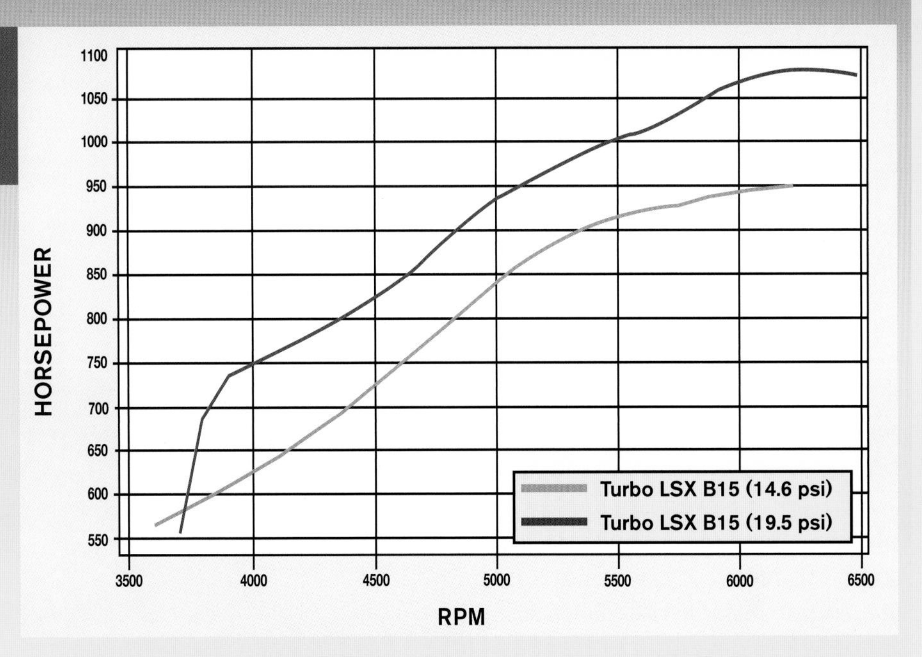

Effect of Boost on a Turbo LSX B15 (14.6 vs 19.5 psi) (Torque)

Turbo LSX B15 (14.6 psi): 891 ft-lbs @ 5,200 rpm
Turbo LSX B15 (19.5 psi): 979 ft-lbs @ 5,000 rpm
Largest Gain: 152 ft-lbs @ 4,100 rpm

The manual waste-gate controller did not allow me to dial in the boost curve precisely through the entire rev range. Since I was getting close to the maximum output of the single Precision turbo and the back pressure was escalating, the boost curve was not consistent through the rev range and therefore, the LSX with 19.5 psi spiked above 979 ft-lbs early in the run. Despite this fact, the change in boost offered some serious torque gains.

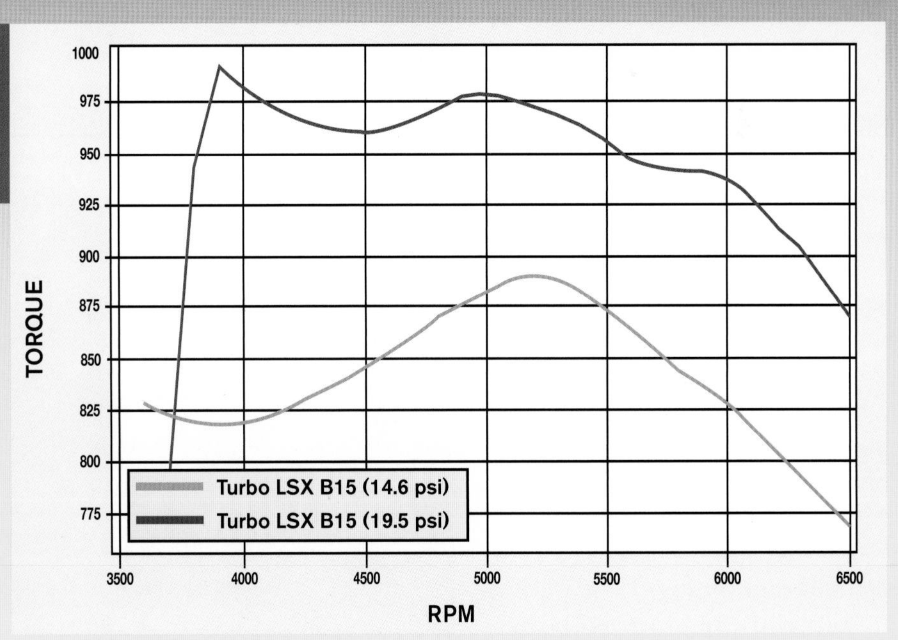

Test 6: 4.8 LS3 Hybrid: NA vs Single Turbo at 9.8 psi

This test proves that turbo boost can be applied to any LS3 combination, including a short-stroke engine. This short-stroke engine was used extensively with both cathedral and rectangular-port heads and ran safely to 8,000 rpm with a hydraulic roller cam (see Chapter 3). The hybrid was the result of combining a 4.8 crank with a big-bore, LS3 aluminum block. Gandrud Chevrolet supplied the new GM block and stuffed it with forged internals from Lunati and JE. The short-stroke LS3 was topped for this test with TFS Gen X 255 LS3 heads that flowed more than 380 cfm. I liked the fact that the TFS heads featured even smaller port volumes than the stock heads (good for reduced displacement). The engine also featured an ATI dampener and complete Moroso oiling system (both critical at high RPM). Tuning came from a Holley EFI management system controlling 120-pound injectors.

For this test, the LS3 hybrid was equipped with a factory LS9 cam. (For details on how much a turbo cam might be worth, see Test 3 in this chapter.) The aluminum test engine was equipped with a single 76-mm Precision turbo capable of easily exceeding the intended power level for this comparison.

I ran the test engine in NA trim before subjecting it to boost. The hybrid produced 512 hp at 6,900 rpm and 415 ft-lbs of torque at 6,200 rpm. Credit the lack of displacement and Holley Hi-Ram intake for the elevated engine speeds (despite the mild cam timing).

Run with the single turbo system pushing out 8.5 psi at the power peak, the turbo hybrid produced 701 hp at 6,500 rpm and 598 ft-lbs of torque at 5,500 rpm. Given the forged internals and 1,200-hp capability of the turbo, there was plenty more left in the combination, but this 700-hp turbo LS idled like a stocker and would provide thousands of trouble-free miles at this boost level.

Tuning for the short-stroke turbo engine was provided by a Holley Dominator EFI system. The Holley was used to control timing and fuel from the 83-pound Holley injectors.

The test engine was an LS3 aluminum block stuffed with a 4.8 crank, forged Lunati rods, and JE pistons. I topped it with a set of TFS Gen X 255 heads and Holley Hi-Ram intake. Note also the use of a Moroso oil pan and ATI dampener.

4.8 LS3 Hybrid: NA vs Single Turbo at 9.8 psi (Horsepower)

Short-Stroke LS3 (NA): 512 hp @ 6,900 rpm
Short-Stroke Turbo LS3 (8.5 psi):
701 hp @ 6,500 rpm
Largest Gain: 197 hp @ 6,500 rpm

The NA short-stroke LS3 was no slouch at 512 hp, but things really got serious once I added boost. Running a peak of 8.5 psi, the turbo LS produced 701 hp, with plenty left in both the engine and turbo.

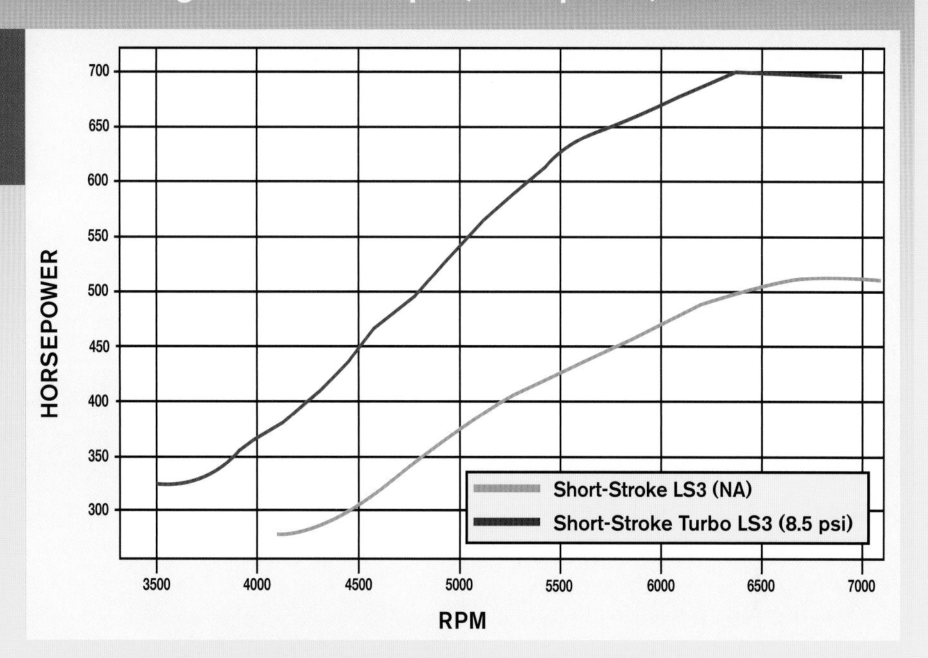

4.8 LS3 Hybrid: NA vs Single Turbo at 9.8 psi (Torque)

Short-Stroke LS3 (NA): 415 ft-lbs @ 6,200 rpm
Short-Stroke Turbo LS3 (8.5 psi):
598 ft-lbs @ 5,500 rpm
Largest Gain: 192 ft-lb 5,500 rpm

Despite the reduced displacement (compared to an LS3) and the use of a rather large 76-mm turbo from Precision, the boost response was very good and so were the torque gains. The boost increased torque production by as much as 192 ft-lbs, with consistent gains through the rev range.

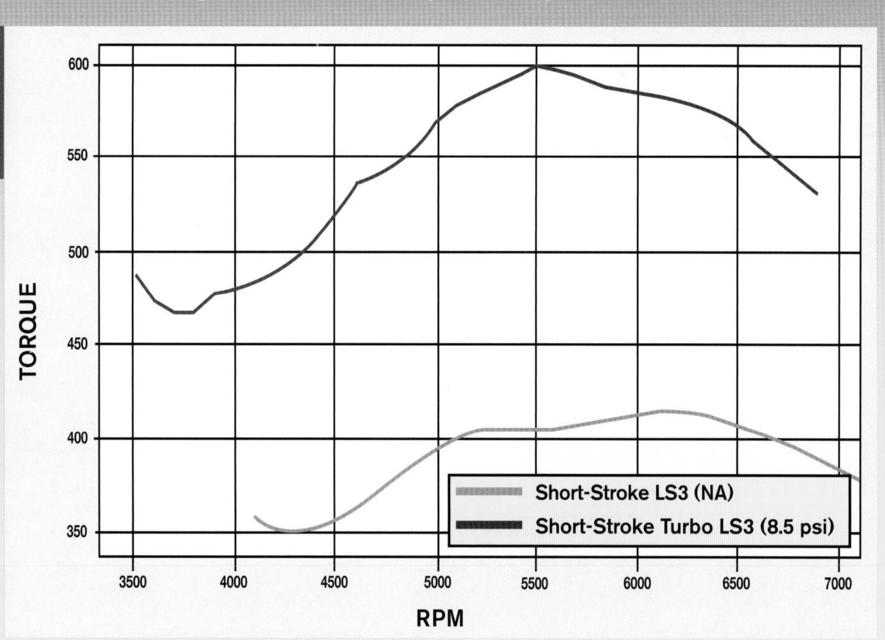

Test 7: Turbo LS: Effect of Snow Water/Methanol Injection

You may be wondering why I decided to include a test on water/methanol injection in this chapter. In reality, water/methanol injection is a form of intercooling, which should be employed on any turbocharged (or supercharged) engine, almost regardless of the boost level.

Although the Snow Boost Cooler water/methanol injection does not provide additional power in the same way as (say) nitrous oxide, it does dramatically decrease the inlet air temperature. This combined with the extra octane offered in the methanol portion of the mixture allows you to be more aggressive on the ignition timing, air/fuel ratio, and/or boost pressure to improve the power output.

The higher the intake charge temperature, the higher the risk of the fuel self-igniting. Having the mixture ignite prior to the piston being in the proper position can result in the expanding mixture working against the upward moving piston. At the very least, this has a detrimental effect on power production; at the very most, it can cause catastrophic engine failure.

In addition to minimizing the chance of detonation, a cooler inlet charge temperature can also provide additional power thanks to the increase in air density. Cooler air has more oxygen molecules per volume, so getting cool air to your engine should be considered mandatory. This is especially true of turbocharged (and supercharged) engines, where the compression (boost) has an elevated charge temperature well above ambient. In the case of the turbocharged LS running 8.5 psi of boost, the inlets air temperatures exiting the turbo exceeded 185 degrees. The Snow water meth system dropped the air temperatures to 95 degrees and improved power by as much as 29 hp and 32 ft-lbs of torque. Water/methanol injection is especially important when running pump gas because it allows for changes in timing and air/fuel that dramatically improve power (see Test 1).

This system featured dual nozzles, but I employed only minimal pressure and the smallest nozzle sizes on the low-boost LS.

The Boost Cooler from Snow Performance can be thought of as chemical intercooling. The injection of a water/methanol mixture dramatically cools the intake air temperature while decreasing the chance of detonation.

Turbo LS: Effect of Snow Water/Methanol Injection (Fluid Air Temperature)

No Snow: 185 degrees @ 6,800 rpm,
Snow Boost Cooler: 87 degrees @ 6,250 rpm

Adding the Snow Boost Cooler water/methanol injection to the turbo LS dramatically decreased the inlet air temperature. The super-cooling system dropped the charge temps from 185 to 94 degrees at 8.5 psi of boost. This drop in temperature not only increases the power output, but allows for additional timing and changes in air/fuel to further improve power production.

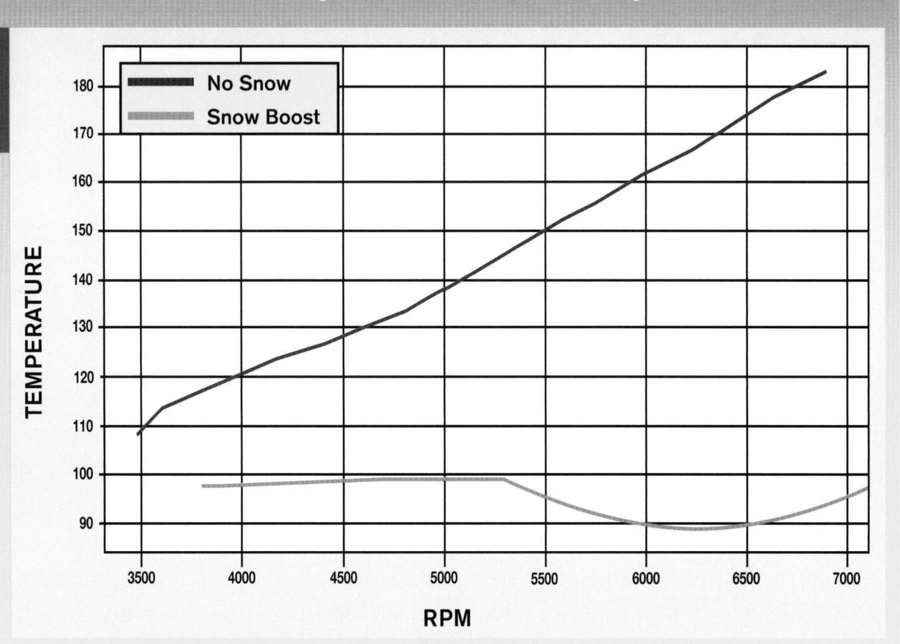

Turbo LS: Effect of Snow Water/Methanol Injection (Horsepower and Torque)

No Snow: 705 hp @ 6,800 rpm
No Snow: 584 ft-lbs @ 5,600 rpm
Snow Boost Cooler: 735 hp @ 6,600 rpm
Snow Boost Cooler: 606 ft-lbs @ 5,500 rpm

Gains offered by the combination of the cooling effect and changes in timing were substantial on this turbo LS application. The changes in charge temperature would be even greater on higher boost applications. The Snow system netted an additional 29 hp and 32 ft-lbs of torque on this turbo LS.

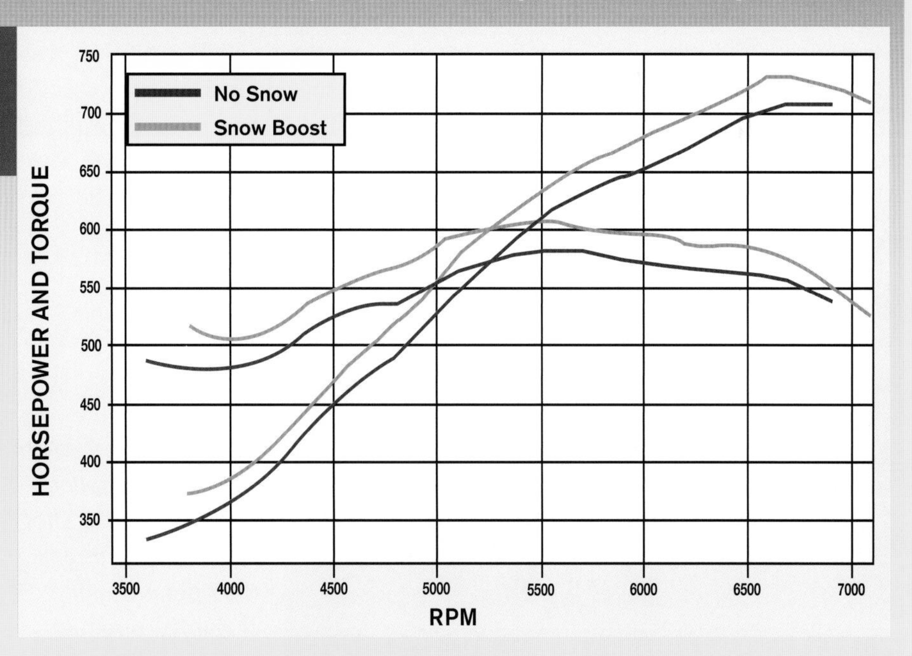

CHAPTER 7

NITROUS OXIDE

When it comes to bang for the buck, nothing compares to nitrous oxide. Toss in the fact that it is easy to install (and conceal) and the power output is adjustable (as with forced induction). It's easy to see why nitrous oxide is all the rage among street racers and enthusiasts. Short of a well-prepared turbo or blower, nothing runs harder than an LS engine on the juice.

Nitrous oxide offers a number of benefits, including the ability to adjust the available power level. Much like cranking up the boost pressure on a turbo, jet changes on a nitrous system allow you to literally dial in the extra power. However, there is a limit to the amount of nitrous that can be added, something usually dictated by the strength of the internal components and the original power output of the engine.

In addition to the adjustable power, street racers like nitrous because it can be easily hidden. Of course, it doesn't take a genius to figure out that a stock 2010 Camaro was sporting something more than the stock when it knocks out consistent 11s at the track. Further improving upon the adjustable power and concealment is the cost. Compared on the basis of available power gains, nitrous offers far and away the best bang for the performance buck.

Nothing wakes up an LS3 or LS7 application like a small shot of nitrous. Combine the right amount of nitrous and fuel through a single (or multiple) fogger nozzle(s) and you have instant horsepower.

As you know, nitrous oxide is not a fuel, but rather it's an oxidizer. Despite the automotive infernos depicted in movies such as *The Fast and the Furious*, nitrous oxide does not burn nor is it likely to incinerate a car. You could literally open the bottle of nitrous and touch a match to the spray and the only thing that would happen is that the match would go out. No thunderous explosions, no massive fire balls, just an anticlimactic wisp of smoke as the flame is extinguished by the high-pressure, ice-cold stream of nitrous oxide.

If nitrous oxide doesn't burn, then how does it increase the power output of the engine? The answer is simple: Nitrous oxide adds power by releasing free oxygen molecules contained in the compound. Because oxygen molecules are a key ingredient in power production (the more oxygen present, the greater the power potential), the release of these oxygen molecules adds to the power potential of the engine. More nitrous equals more free oxygen molecules, which in turn equals more power.

There is, however, a limit to the amount of nitrous and the number of attending free oxygen molecules that can be added to any combination. While most stock engines, even those equipped with cast or hypereutectic pistons, withstand an increase of 40 to 50 percent (depending on the original power output and displacement), adding more power brings the strength of the internal components into play.

Building a high-horsepower nitrous engine is not much different than building a high-horsepower turbocharged or supercharged engine. Short-blocks typically include forged rods, cranks, and pistons, with MLS head gaskets, head studs, and possibly O-ringing the block. Nitrous and forced induction engines do, however, differ in their cam timing and cylinder head porting. Nitrous engines prefer big port volumes and a lot of exhaust flow because the nitrous adds all the necessary intake oxygen molecules. All those extra oxygen molecules must now be allowed to escape, thus the need for greater exhaust port flow and wilder exhaust cam timing.

Adding power through nitrous is different than adding the same amount of power through forced induction. Sure, both add an easy 75, 100, or even 150 hp (or more) to an average LS engine, but how they add the power differs. Both forced induction and nitrous increase the amount of oxygen molecules available to produce power. Forced induction does so by increasing the mass flow of air. Pressurizing the air increases the mass flow, thus force-feeding the engine more air than it could ingest on its own or otherwise in NA form.

The unfortunate side effect of the pressurization of air (boost) is that the pressure causes heat. Turbochargers and superchargers heat the inlet air, something not desirable from either a power (less oxygen molecules per volume) or a detonation threshold standpoint. The hotter the air, the easier it is to ignite. In some cases, the heated inlet air can self-ignite before the spark plug initiates the burn. The result is an expansion of the air/fuel mixture while the piston is still on its way to TDC. As a result, the expanding gases resist the upward moving piston. The result of this struggle is sometimes not very pretty. The same thing can happen with excessive ignition advance.

Nitrous, on the other hand, does not resort to pressurizing the inlet air, but rather the extra oxygen molecules are carried in the pressurized compound. Once delivered to the inlet tract from a pressurized bottle, the liquid nitrous quickly turns into a gas. This liquid-to-gas vaporization requires an input of energy; in this case the energy is heat. The vaporization of the liquid nitrous absorbs heat from the surrounding inlet air, desirable in any performance application (especially a turbo or supercharged engine).

Although you associate heat with boiling (for example, water turning from a liquid to a gas), the vaporized nitrous does not produce heat (at least not to the inlet air). Although vaporized, the temperature of the nitrous oxide is still at or near minus 129 degrees F (the boiling point of nitrous oxide). Mixing the inlet air with a gas that is even that cold still provides a dramatic cooling effect. This double cooling reduces the chance of detonation and increases the density of the inlet air. Denser air equals more oxygen molecules, which in turn (potentially) creates more power.

Nitrous kits can be pretty elaborate but most feature the components illustrated in this NOS kit. The components include a bottle, solenoids, fogger or plate, jetting and arming, and activation switches.

Nitrous oxide (such as in this Zex kit) can also be used on carbureted LS applications with the Perimeter Plate system. This system was designed to evenly distribute the fuel and nitrous to all cylinders to maximize (safe) power.

Test 1: Modified LS3: NA vs Nitrous Using 100- and 150-hp Shots

The great thing about nitrous oxide is that it can be added to any engine, but big doses of nitrous are best used on modified engines designed to take the abuse. After all, nitrous kits can easily increase the power output of an NA engine by 100 hp or more.

This test was run on a modified LS3. Starting with a GM LS3 crate engine from Gandrud Chevrolet, the short-block was augmented with a set of CP forged, flat-top pistons and 6.125-inch connecting rods. The piston and rod upgrade were combined with the stock crank (more than strong enough for this level of nitrous). I retained the stock LS3 heads, but they were treated to a valvespring upgrade from BTR. Comp Cams supplied a 459 cam for this test that offered a .617/.624-inch lift split, 231/239-degree duration split, and 113-degree LSA. The stock heads were retained using Fel Pro MLS head gaskets and ARP head studs. This test relied on the stock LS3 intake, but I swapped on a manual 90-mm throttle body to replace the factory drive-by-wire unit. Run with long-tube headers, a Meziere electric water pump, and FAST XFI management system, the modified LS3 produced 552 hp, and 520 ft-lbs of torque.

To illustrate the power gains offered by nitrous oxide, I selected one of the affordable Sniper Kits from NOS. Because one of the benefits of nitrous is the ability to easily increase power with a simple jet change, I decided to run two different power levels on this engine. The Sniped Universal EFI kit featured a 10-pound bottle, two solenoids, and a single fogger nozzle designed to inject the nitrous and fuel together. Because nitrous oxide is an oxidizer, extra fuel is a critical component to add to the extra oxygen molecules supplied by the compound. Using the supplied jetting, I set up the system to provide an extra 100 hp. Activating the nitrous at 4,300 rpm resulted in a jump in power to 687 hp and 684 ft-lbs of torque. The Sniper system provided a nice, smooth power curve. After the success of the 100-hp shot, I stepped up to 150-hp jetting and was immediately rewarded with 732 hp and 730 ft-lbs of torque. I made sure to heat the bottle properly to ensure adequate bottle pressure prior to testing.

Because the stock valvesprings did not accept the available cam lift, I replaced them with a dual-valvespring upgrade from BTR.

In addition to forged rods and pistons, the LS3 also received a Comp 459 cam upgrade.

Modified LS3: NA vs Nitrous Using 100- and 150-hp Shots (Horsepower)

Modified LS3: 552 hp @ 6,100 rpm
Nitrous Modified LS3 (100-hp shot): 687 hp @ 6,000 rpm
Nitrous Modified LS3 (150-hp shot): 732 hp @ 6,000 rpm
Largest Gain: 189 hp @ 5,800 rpm

The Sniper universal wet EFI system offered plenty of bang for the buck. Applied to this modified LS3, the nitrous system increased the power output from 551 to 687 hp using the 100-hp jetting. This increased to 732 hp with the 150-hp jetting. Thanks to the forged internals from CP and Carillo, I felt confident adding this much power to the LS3.

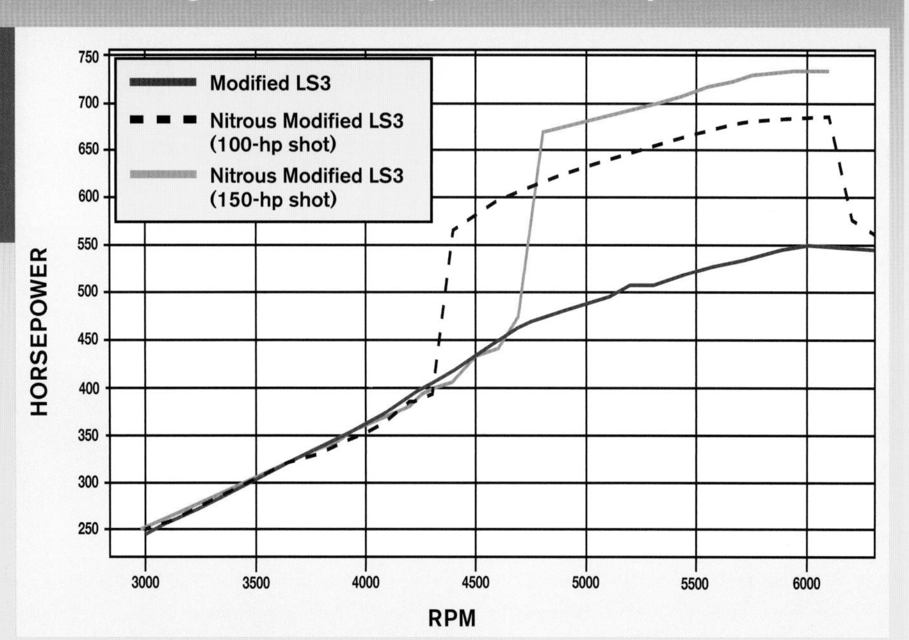

Modified LS3: NA vs Nitrous Using 100- and 150-hp Shots (Torque)

Modified LS3: 520 ft-lbs @ 4,800 rpm
Nitrous Modified LS3 (100-hp shot): 684 ft-lbs @ 4,600
Nitrous Modified LS3 (150-hp shot): 730 ft-lbs @ 4,800 rpm
Largest Gain: 210 ft-lbs @ 4,800 rpm

The torque curves illustrate how easy it was to increase power on the nitrous-injected LS3. Simple by changing jets on the Sniper system, I was able to increase torque production first from 520 ft-lbs to 684 ft-lbs, then up to 730 ft-lbs with the 150-hp jetting. Note the 300-rpm earlier activation (4,300 versus 4,600) on the 100-hp shot.

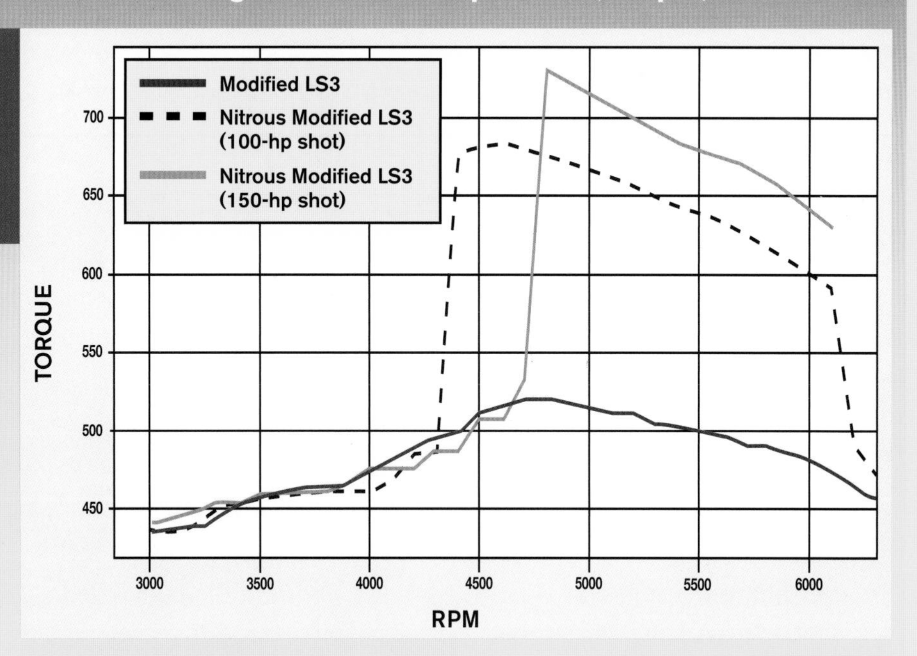

Test 2: 408 Hybrid Stroker: NA vs Zex Wet EFI Nitrous Using a 100-hp Shot

Aluminum LS3 engines are expensive and more difficult to come by than iron 6.0 truck blocks, and as a result, building your own 6.0 LS3 hybrid has become common practice. This works especially well if you bore and stroke the 6.0 to 408 ci as with this test engine.

The 6.0 iron block was first bored .030 over then treated to a stroker assembly that included a Scat forged steel crank and 6.125-inch rods combined with JE forged (asymmetrical) pistons. Finishing the stroker was a Comp cam (PN 277LrHR13) that offered a .614/621-inch lift split, 227/235-degree duration split, and 113-degree LSA. Making this a hybrid was the fact that I topped off the .030-over iron block with a set of as-cast LS3 heads treated to a Comp beehive valvespring (PN 26918) upgrade. The engine was run with a stock LS3 intake, FAST injectors, and a Holley HP management system. As always, I replaced the drive-by-wire throttle body with a 90-mm manual version. Equipped as such, the NA LS3 hybrid produced 577 hp and 526 ft-lbs of torque.

After running the NA hybrid stroker, I installed the Zex Wet EFI nitrous kit. The cool thing about the Zex kit is that not only was it adjustable with different jetting, but the single fogger nozzle could be applied to just about any fuel-injected application. The kit included the purple 10-pound bottle, a digital controller, and a single fogger nozzle designed to combine the nitrous and fuel before injecting it into the engine.

After filling the bottle at Westech Performance, I hooked up the system, went through the WOT learn procedure for the throttle position sensor (TPS), then purged the system. With plenty of bottle pressure, I activated the Zex nitrous at 4,500 rpm, and the power jumped immediately. It soon settled in with gains that exceeded 100 hp. Credit the extra power offered by the kit to tuning and bottle pressure above 1,000 psi. Increased bottle pressure works like a larger nitrous jet, and this Zex kit worked amazingly well on this hybrid stroker.

The test engine started as a 6.0, but the iron block was upgraded with a 4.0-inch SCAT stroker crank and K1 6.125-inch rods.

JE supplied the necessary forged flat-top pistons for the LS3-headed 6.0 stroker.

408 Hybrid Stroker: NA vs Zex Wet EFI Nitrous Using a 100-hp Shot (Horsepower)

NA 408 LS3 Stroker: 577 hp @ 6,400 rpm
Zex Nitrous 408 LS3 Stroker: 702 hp @ 6,100 rpm
Largest Gain: 121 hp @ 6,400 rpm

Nothing adds instant power like a good nitrous system. The Zex Wet EFI kit offered impressive power gain, upping the power output of the 408 hybrid stroker from 577 hp to 702 hp using 100-hp jetting. I did take the necessary steps to ensure proper nitrous pressure and delivery by preheating the bottle.

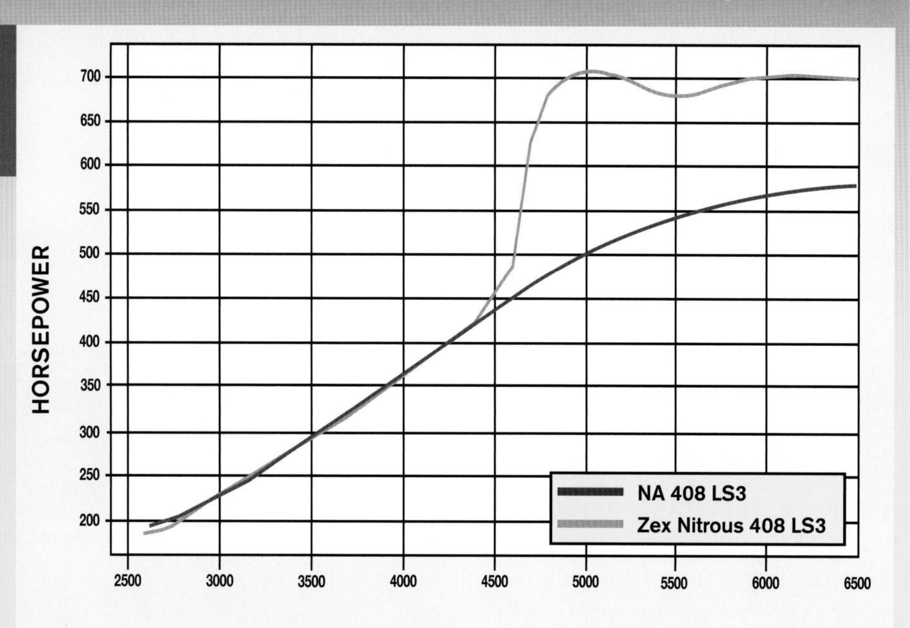

408 Hybrid Stroker: NA vs Zex Wet EFI Nitrous Using a 100-hp Shot (Torque)

NA 408 LS3 Stroker: 535 ft-lbs @ 5,200 rpm
Zex Nitrous 408 LS3 Stroker: 827 ft-lbs @ 5,400 rpm
Largest Gain: 211 ft-lbs @ 4,900 rpm

It is not unusual to see a massive torque gain on the initial activation of a nitrous system. The Zex kit added more than 200 ft-lbs to the stroker engine at 4,900 rpm but settled in to a more realistic torque gain thereafter. Remember, the lower you activate the nitrous, the more torque you gain, but care must be taken not to become too greedy.

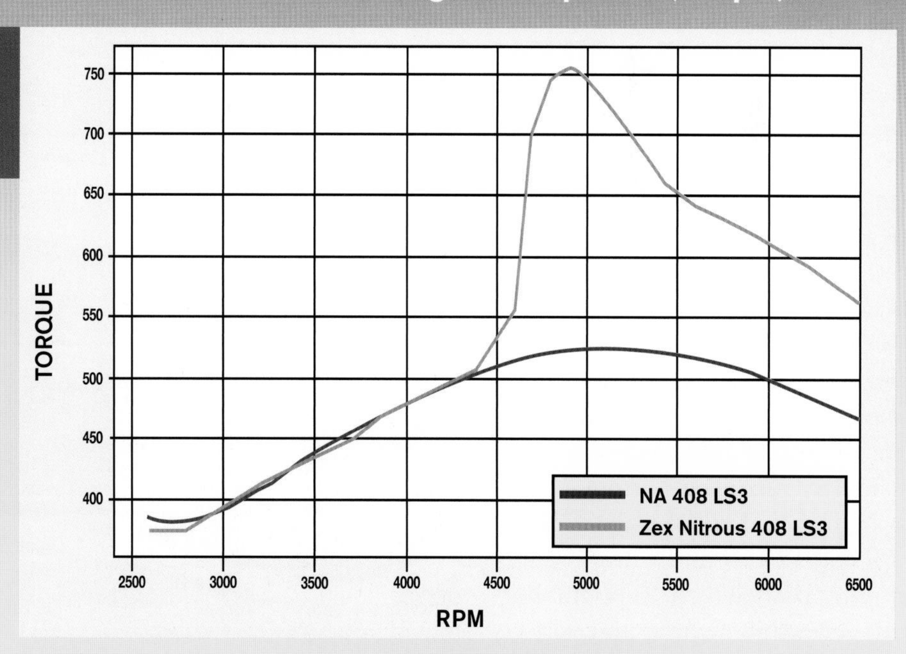

Test 3: Mild LS3: NA vs Nitrous Works Using a 100-hp Shot

I ran this test to illustrate that it is possible to not only add nitrous to any LS3 or LS7 combination (even a stock one), but do so safely. Forget the horror stories and explosions shown on TV, movies, or YouTube, nitrous is not even flammable, nor does it hurt stock pistons or rods if tuned properly. Nitrous is simply an oxidizer that adds extra oxygen molecules. Extra oxygen means extra power, but the oxygen must have additional fuel to support the burn. Combine the two properly and inject into the engine and that's when the magic starts to happen.

To illustrate how well nitrous companies have perfected their systems, I ran nitrous on an LS3 equipped with stock internals, meaning powdered-metal rods and cast pistons. The only upgrades made to the LS3 were a BTR cam and valvesprings. Run with headers, a Meziere electric water pump, and Holley HP management system, the cam-only LS3 produced 544 hp and 514 ft-lbs of torque. (See Chapter 3 for some serious cam-only information from BTR.).

As much as I love how well an LS3 responds to more aggressive cam timing, it responds even better to nitrous oxide. This includes an old, out-of-date system no longer available. Rummaging through the cabinets at Westech Performance, I ran across an old Nitrous Works kit from Barry Grant. No longer available, the system was complete and ready to run. The Nitrous Works system included the usual array of solenoids, lines, and a single fogger nozzle, along with a 10-pound bottle, activation switch, and various hoses and fittings.

To test the system, I installed jetting to provide a 100-hp increase to the cam-only LS3. Despite not being available for some time, all the components worked well and the 100-hp jetting increased the power output of the LS3 from 544 to 660 hp. The torque gains were equally impressive; the nitrous increased torque production from 514 to 655 ft-lbs. Even on a near-stock engine, nitrous oxide works wonders.

The power output of the GM LS3 crate engine was increased by replacing the factory cam with a higher lift cam from BTR.

The cam swap necessitated replacement of the stock LS3 valvesprings. BTR supplied a set of double springs to accommodate the higher-lift and increased engine speed of the new cam.

Mild LS3: NA vs Nitrous Works Using a 100-hp Shot (Horsepower)

NA LS3: 546 hp @ 6,000 rpm
Nitrous Works LS3: 660 hp @ 6,100 rpm
Largest Gain: 122 hp @ 5,500 rpm

Starting with a GM LS3 crate engine from Gandrud Chevrolet, I installed a mild BTR cam (and springs), then an old Barry Grant Nitrous system that I had laying around. Running jetting to supply an extra 100 hp, the single fogger system increased the power output of the mild LS3 from 554 hp to 660 hp. Nitrous offered nice, smooth, consistent gains.

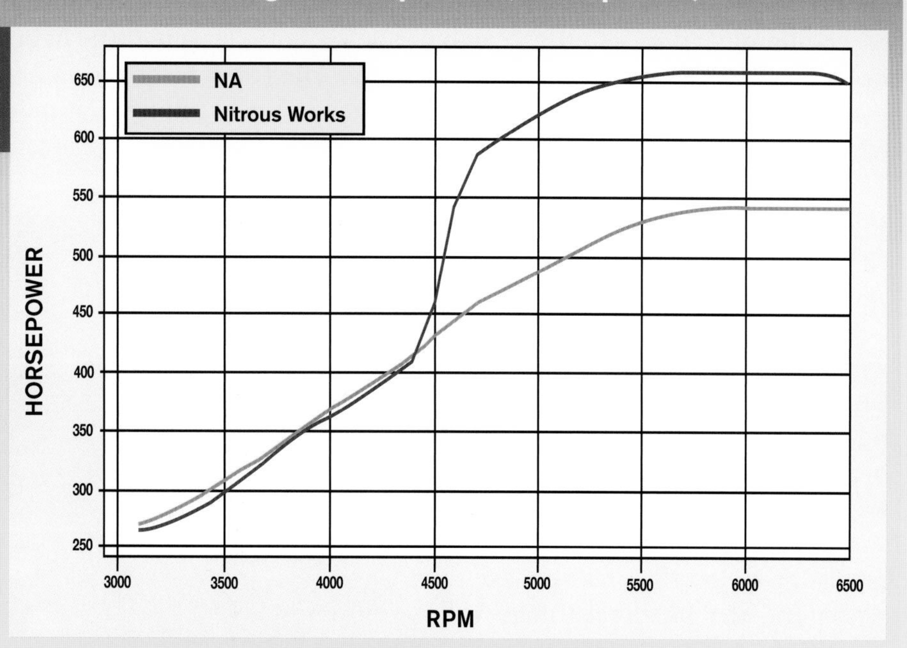

Mild LS3: NA vs Nitrous Works Using a 100-hp Shot (Torque)

NA LS3: 514 ft-lbs @ 4,800 rpm
Nitrous Works LS3: 655 ft-lbs @ 4,800 rpm
Largest Gain: 141 ft-lbs @ 4,800 rpm

Activation of the Nitrous Works nitrous system at 4,500 rpm resulted in impressive torque gains. I made sure to retard the timing by 4 degrees to eliminate any chance of detonation. Nitrous can transform a mild LS3 into a wild one with one push of the button.

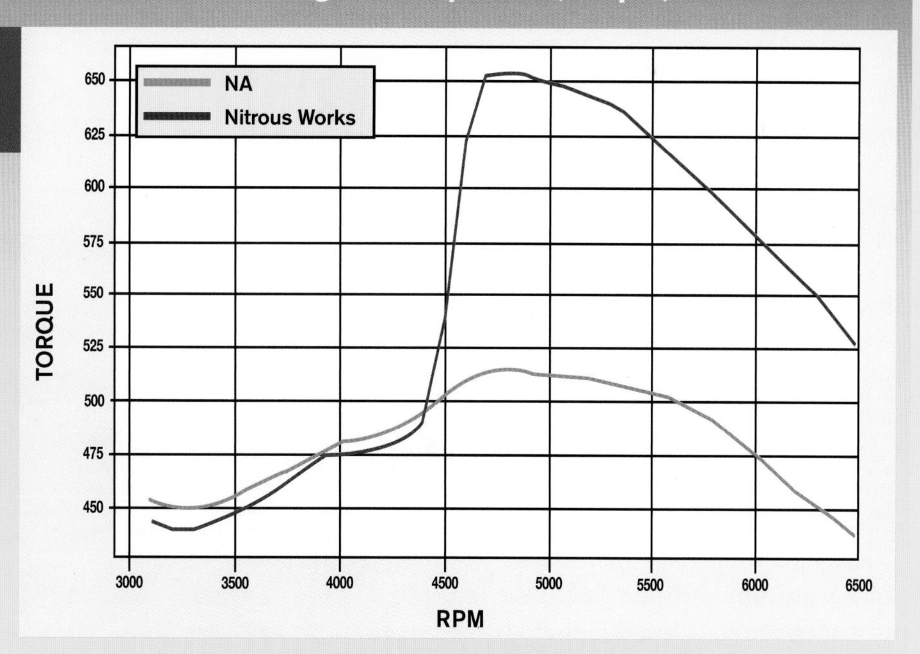

Test 4: LS3-Headed 6.0L: NA vs Nitrous Using a 100-hp Shot

What is the next best thing to the LS3? Based on numbers alone, the obvious answer is an LS2, but this test engine offered an effective combination of both. Looking at the factory power ratings, the LS3 offered both more displacement and better head flow than the LS2, but that doesn't mean an LS2 can't make power like an LS3. All you have to do is replace the LS2 heads with the high-flow LS3 heads and intake manifold.

Although the swap can be performed on a stock short-block, this one was augmented with JE pistons and K1 rods (using the stock crank). Since the heads interchanged, the swap was easy, but replacing the LS2 heads with the LS3 heads required installation of the LS3 offset intake rockers and the LS3 intake manifold. The LS3 throttle body was also ditched in favor of a manual version and the 6.0 hybrid was run with a Holley HP EFI system, long-tube headers, and a Meziere electric water pump. Run with Lucas oil, the 6.0 stroker produced 493 hp and 461 ft-lbs of torque.

With a short-block at the ready, I decided it was plenty stout enough for some nitrous testing. For this 6.0/LS3 hybrid, I chose the NOS No Spray Bar Nitrous Plate system. Designed to sandwich between the throttle body and intake manifold, the system combined the fuel and nitrous to fire it directly into the awaiting intake manifold. The system was adjustable from 100 to 150 hp, but since the system employed Cheater solenoids, optional jetting could take the system to 250 hp. For the 6.0 hybrid, I chose to stick with mild jetting to provide 100 hp. After installation of the system, I heated the bottle to get the pressure above 900 psi.

After pushing the activation button, the power output of the hybrid jumped from 493 hp and 461 ft-lbs of torque to 607 hp and 630 ft-lbs of torque. Any LS that pumps out more than 600 hp is a force to be reckoned with on the street.

The aluminum 6.0 LS2 block was equipped with LS3 heads to produce a hybrid. The build included JE forged pistons, K1 rods, and a stock crank. ARP head studs and Fel Pro MLS head gaskets were also used.

Tuning for the injected 6.0 LS3 came from a Holley Dominator. Dialing in both air/fuel and timing is critical when using nitrous. The Dominator was used to retard the ignition timing by 4 degrees with a 100-hp shot.

LS3-Headed 6.0L: NA vs Nitrous Using a 100-hp Shot (Horsepower)

NA 6.0 LS3: 493 hp @ 6,200 rpm
NOS 6.0 LS3: 607 hp @ 6,100 rpm
Largest Gain: 148 hp @ 5,100 rpm

How do you not love a system that adds an easy 100 (or more) hp, regardless of what engine you are running? Adding the NOS system to the LS3-headed 6.0 increased the power output from 493 to 607 hp.

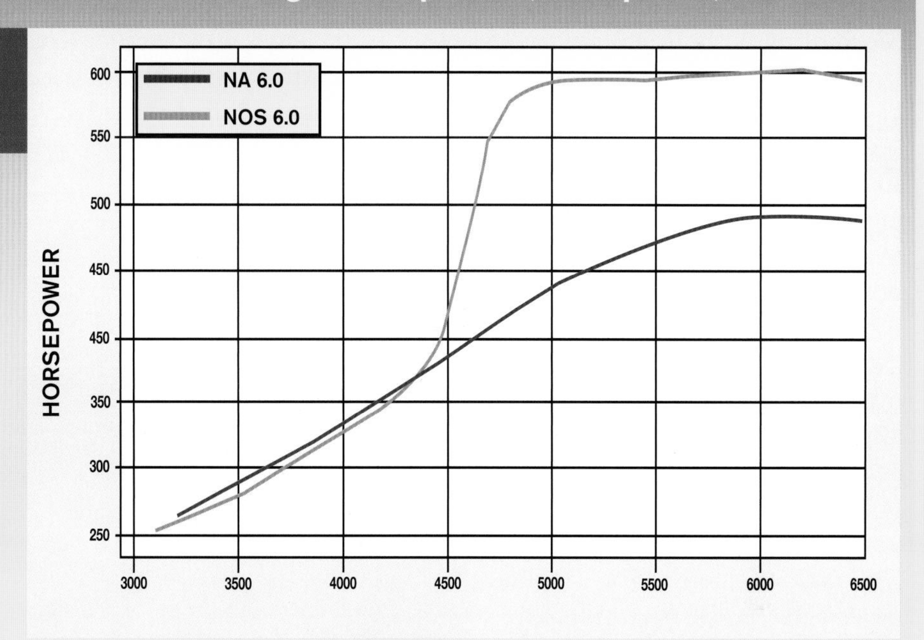

LS3-Headed 6.0L: NA vs Nitrous Using a 100-hp Shot (Torque)

NA 376 LSX B15: 461 ft-lbs @ 5,000 rpm
NOS 6.0 LS3: 631 ft-lbs @ 4,900 rpm
Largest Gain: 177 ft-lbs @ 4,800 rpm

Activating the NOS nitrous system at 4,000 rpm brought huge torque gains between 4,500 and 5,000 rpm. Because of the relationship between horsepower, torque, and RPM, simple math tells you that if you activate the system below 5,252 rpm, the torque gains will exceed the horsepower gains, meaning you will get considerably more than 100 ft-lbs if you add a 100-hp shot. The 100-hp shot on this LS3-headed 6.0 offered as much as 177 ft-lbs.

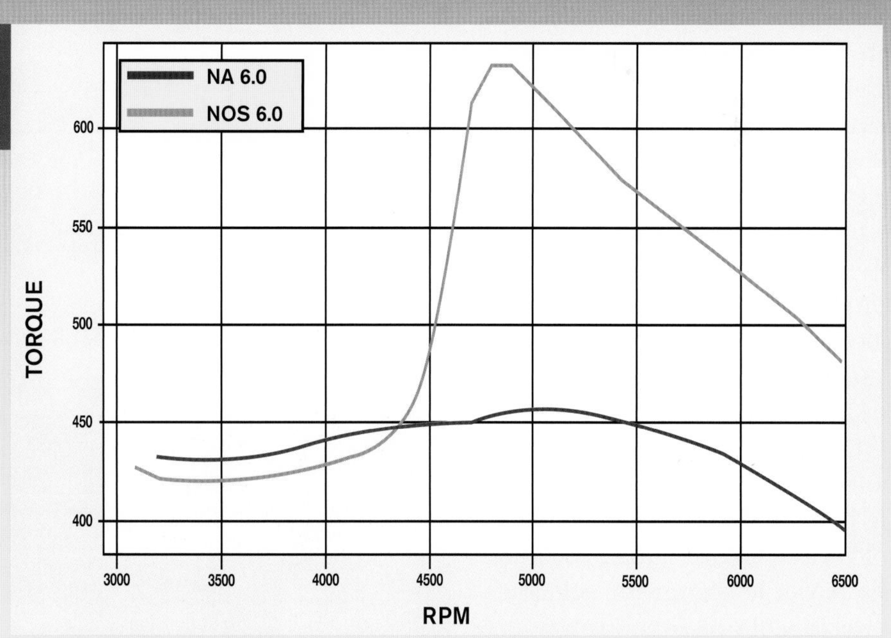

Test 5: LS3 Stroker: NA vs Nitrous Using a 150-hp Shot

Things really get serious when you combine cubic inches and nitrous oxide. Case in point: this 417 stroker LS3. Starting with an aluminum LS3 block from Gandrud Chevrolet, I installed a 4.0-inch forged-steel stroker crank and 6.125-inch connecting rods from Lunati. Completing the rotating assembly was a set of CP forged, flat-top pistons and Total Seal rings. The idea was to build a stout, NA combination to which I would add a healthy dose of nitrous. The 417 was completed using a set of CNC-ported L92 heads equipped with a dual-valvespring upgrade from BTR, as well as a camshaft for the stroker.

Designed for a 400-inch stroker, the BTR Cam offered a .631/.595-inch lift split, 251/266-degree duration split, and 112-degree +4 LSA. The engine was run with the stock LS3 intake, stock rockers, and Hooker long-tube headers using a Holley HP management system.

Before adding the NOS billet plate LS nitrous system, the engine was run in NA trim. As with any nitrous system, the power gains add to the existing output. In the case of this LS3 stroker, the engine produced 605 hp at 6,500 rpm and 542 ft-lbs of torque at 5,200 rpm. Torque production exceeded 500 ft-lbs from 4,300 rpm to 6,300 rpm, making for a broad power band.

The NOS billet plate system bolted between the 92-mm throttle body and factory intake manifold. The kit used jetting to adjust the power level, but the solenoids were capable of supporting more than 200 hp. I selected 150-hp jetting for this stroker and retarded the timing by 6 degrees before hitting the button. After activation of the nitrous, the peak power numbers jumped to 779 hp and 799 ft-lbs of torque. Credit the early engagement (4,500 rpm) for big-time torque gains. Run on nitrous, this stroker exceeded 750 hp over a broad range, meaning this would be a serious contender on the track (or street).

Starting with an LS3 block from Gandrud Chevrolet (GM Performance), I added a stroker assembly that included a forged Lunati crank, matching 6.125-inch rods, and CP Pistons.

The 417-inch LS3 stroker was topped with a set of GM CNC-ported L92 heads equipped with a BTR valvespring upgrade.

LS3 Stroker: NA vs Nitrous Using a 150-hp Shot (Horsepower)

NA LS3 Stroker: 605 hp @ 6,500 rpm
Zex Nitrous LS3 Stroker (150-hp shot): 779 hp @ 6,200 rpm
Largest Gain: 184 hp @ 5,000 rpm

The Zex kit improved the power output substantially, adding 174 hp (peak to peak), and there were huge gains from 4,700 to 7,000 rpm. This nitrous-injected stroker offered one sweet power band, pulling strong all the way to 7,000 rpm. Talk about fun on the freeway!

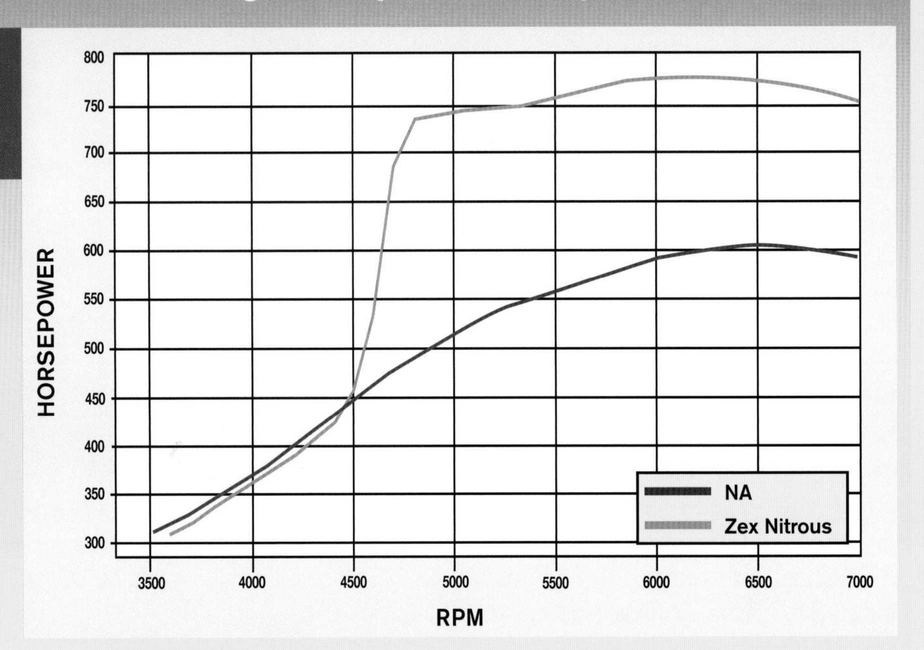

LS3 Stroker: NA vs Nitrous Using a 150-hp Shot (Torque)

NA LS3 Stroker: 542 ft-lbs @ 5,200 rpm
Zex Nitrous LS3 Stroker (150-hp shot): 799 ft-lbs @ 4,800 rpm
Largest Gain: 261 ft-lbs @ 4,800 rpm

Starting with more cubic inches and adding more nitrous is a surefire route to serious performance. This LS3 stroker offered more than 542 ft-lbs of torque, but the output jumped to 799 ft-lbs after activation of the Zex Perimeter Plate system.

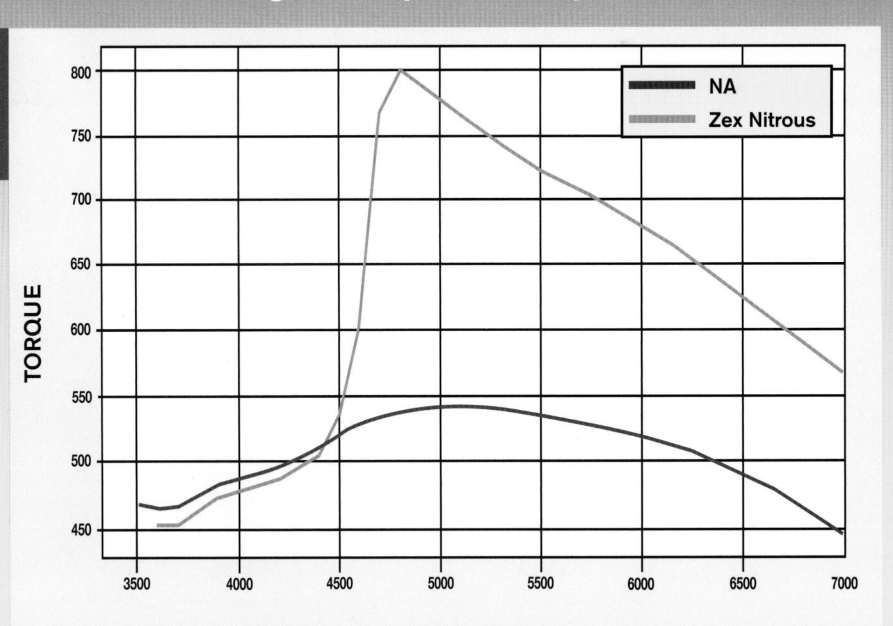

Test 6: Stock LS3: NA vs Zex Nitrous Using a 200-hp Shot

As should be evident by the other tests in this chapter, it is possible to add nitrous to any combination, including a stone-stock LS3 (or LS7) application. To illustrate this fact, I added nitrous to a stock LS3 crate engine from Gandrud Chevrolet, and I added NITROUS. Why the caps, you ask? Because I added not some wimpy 75-, 100-, or even 125-hp shot, but a full 200 hp.

Although this may seem like a lot of juice for a stock engine to ingest, the reality is that a stock LS3 is a pretty impressive customer, offering nearly 500 hp on an engine dyno in the test configuration (headers, open exhaust, and no accessories). When you compare this 200-hp shot to the 493 hp offered by the LS3 crate engine, you see that even the *big* shot of nitrous represented a 40-percent increase in power. This was a lot of nitrous to add to an internally stock engine, but I took the necessary precautions before hitting the button and was rewarded with big power; and the LS3 was ready for more.

To prep for the test, I configured the injected LS3 crate engine with a set of Hooker long-tube LS headers and open exhaust, a Meziere electric water pump, and no air intake feeding the open 92-mm FAST throttle body (stock LS3 intake). Since tuning was critical on a stock engine with this much nitrous, I relied on a FAST XFI system to retard the required 8 degrees of timing upon activation and dial in the air/fuel mixture. As a hedge against detonation, I also filled the fuel tank with 114-octane Rocket Brand race fuel.

Before activation of the Zex single fogger system, I took the liberty of heating the bottle to produce optimum bottle pressure and nitrous flow. After purging the system, I activated the Zex kit on the LS3 crate engine and was rewarded with a jump in power from 493 to 706 hp. The torque gains were even more impressive because the peak grunt jumped from 484 to 737 ft-lbs. Every bit as important was the fact that the LS3 crate engine shrugged off the big hit and was ready for more testing.

The Zex kit run on the LS7 featured a controller that recognized the WOT voltage curve of the throttle position sensor (TPS) to ensure no nitrous would be delivered unless the engine was at WOT.

Bottle pressure is critical to maximize nitrous flow to the Zex controller. I used this pressure gauge with the bottle heater to ensure adequate pressure for each test run on the LS7.

Stock LS3: NA vs Zex Nitrous Using a 200-hp Shot (Horsepower)

Stock NA LS3: 493 hp @ 5,700 rpm
Zex Nitrous LS3: 706 hp @ 5,700 rpm
Largest Gain: 211 hp @ 5,200 rpm

These are the graphs I love to see when running nitrous. The power curve offered by the stock LS3 reached nearly 500 hp, but adding a 200-hp shot of Zex pushed the power output beyond 700 hp. Truth be told, a 200-hp shot on a stock engine might be excessive, but I ran this test on race fuel to eliminate any chance of detonation.

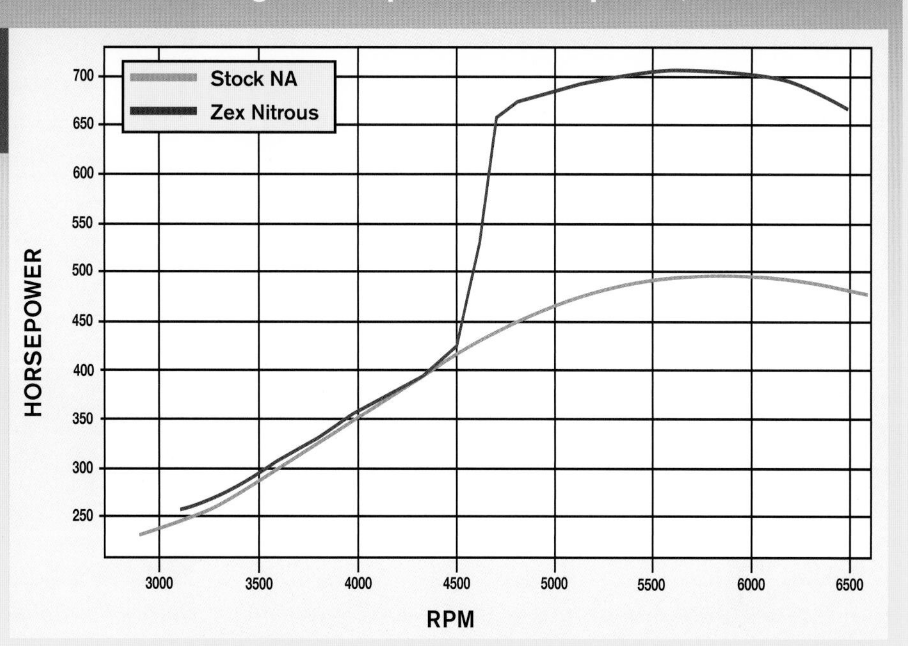

Stock LS3: NA vs Zex Nitrous Using a 200-hp Shot (Torque)

Stock NA LS3: 484 ft-lbs @ 4,600 rpm
Zex Nitrous LS3: 737 ft-lbs @ 4,700 rpm
Largest Gain: 238 ft-lbs 4,800 rpm

The torque curves were equally smooth, with huge gains offered below 5,000 rpm. Care must be taken not to activate the nitrous too early because excessive cylinder pressure can ruin a perfectly good engine. On the stock LS3, torque jumped from 484 to 737 ft-lbs.

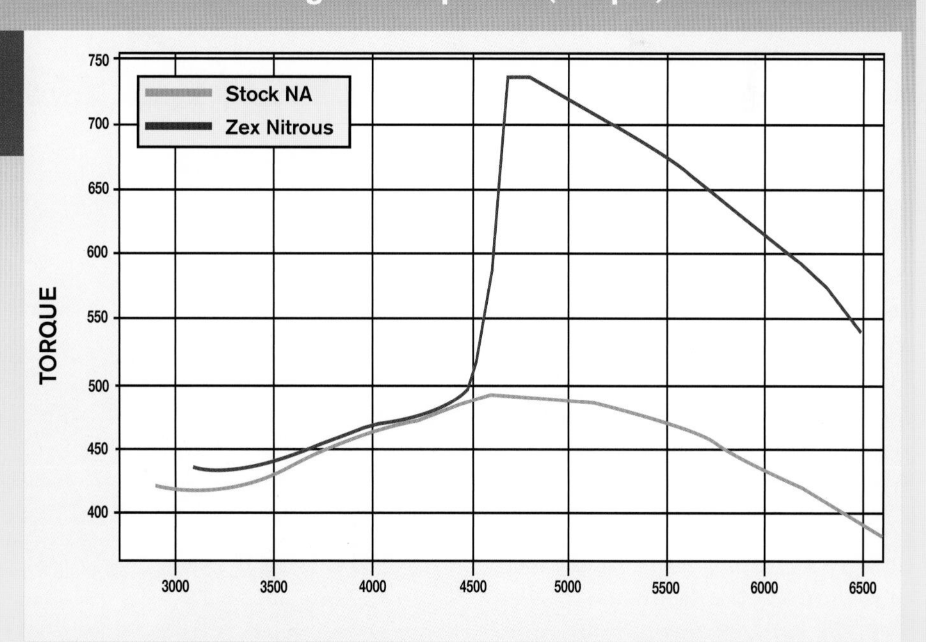

Test 7: Cam-Only LS7: NA vs Nitrous Express Using 125- and 175-hp Shots

As the heavy hitter of the NA LS family, the LS7 has big shoes to fill. Sharing the name with a 12.0:1, rectangular-port, 454 crate engine (not offered in a production vehicle) meant that the modern LS7 better make some serious power. Luckily for LS enthusiasts, the modern 427 commands plenty of respect on the street and on the track.

What's not to love about an all-aluminum LS engine offering 7.0 liters of displacement (427 ci), raised rectangular-port heads, and even a dry-sump oiling system? It took forced induction for the LSA and LS9 powerplants to eventually surpass the LS7 in terms of factory performance, but even the mighty big blocks of the muscle car didn't make as much power as this modern 427. As good as the factory LS7 is, it can always make more power. In this case, more power came from a small shot of nitrous oxide.

Before I switched to the heated water tank to warm the nitrous bottle, I positioned it in front of this space heater (in retrospect, not a good idea, but I kept an eye on it).

Although the 427 LS7 is plenty powerful in stock trim, this engine was equipped with a mild Comp cam and valvespring upgrade. The mild cam helped it produce just over 600 flywheel hp on the engine dyno. Run with the stock LS7 intake, Hooker headers and a Holley HP management system, the cam-only LS7 produced peak numbers of 602 hp and 557 ft-lbs of torque. While the cam peaked at 6,500 rpm, the LS7 pulled strong all the way to 7,000 rpm.

Like most systems, the Nitrous Express kit was adjustable using the supplied jetting. I heated the bottle once more until the pressure exceeded 900 psi, then let it rip with 125-hp jetting. The reward for proper preparation and tuning was a bunch of extra horsepower and torque. The peak numbers jumped to 726 hp and 641 ft-lbs of torque. Things escalated even further after I installed 175-hp jetting. This brought the totals to 784 hp and 721 ft-lbs of torque. The only thing better than an LS7 is one with nitrous.

The LS7 was upgraded with a mild cam from Comp Cams. The factory springs were also upgraded in the process.

Cam-Only LS7: NA vs Nitrous Express Using 125- and 175-hp Shots (Horsepower)

NA Cam-Only LS7: 602 hp @ 6,500 rpm
Zex Cam-Only LS7 (125 hp): 726 hp @ 6,400 rpm
Zex Cam-Only LS7 (175 hp): 784 hp @ 6,400 rpm
Largest Gain: 186 hp @ 6,300 rpm

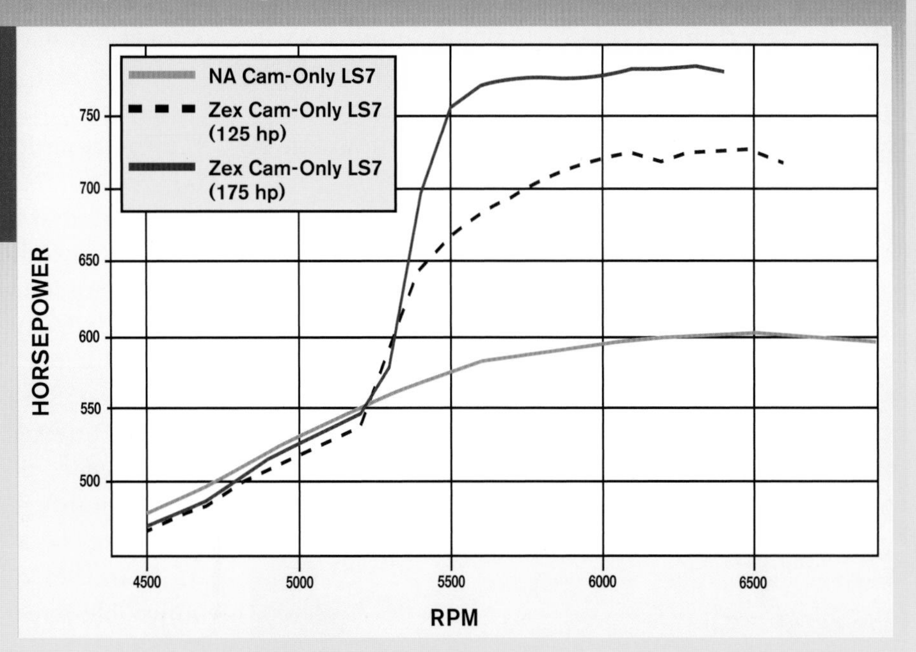

As I often do when testing nitrous oxide on a combination, I ran this cam-only LS7 at different power levels. I started by adding a 125-hp shot then followed it with a slightly larger 175-hp shot. Because the LS7 featured stock rods and pistons, I limited the testing to 175 hp and ran the testing with 100-octane race fuel. Run in NA trim, the LS7 produced 602 hp, but this jumped to 726 hp with the 125-hp jetting, then to 784 hp with the 175-hp jetting.

Cam-Only LS7: NA vs Nitrous Express Using 125- and 175-hp Shots (Torque)

NA Cam-Only LS7: 557 ft-lbs @ 5,000 rpm
Zex Cam-Only LS7 (125 hp): 641 ft-lbs @ 5,600 rpm
Zex Cam-Only LS7 (175 hp): 721 ft-lbs @ 5,500 rpm
Largest Gain: 177 ft-lbs @ 5,600 rpm

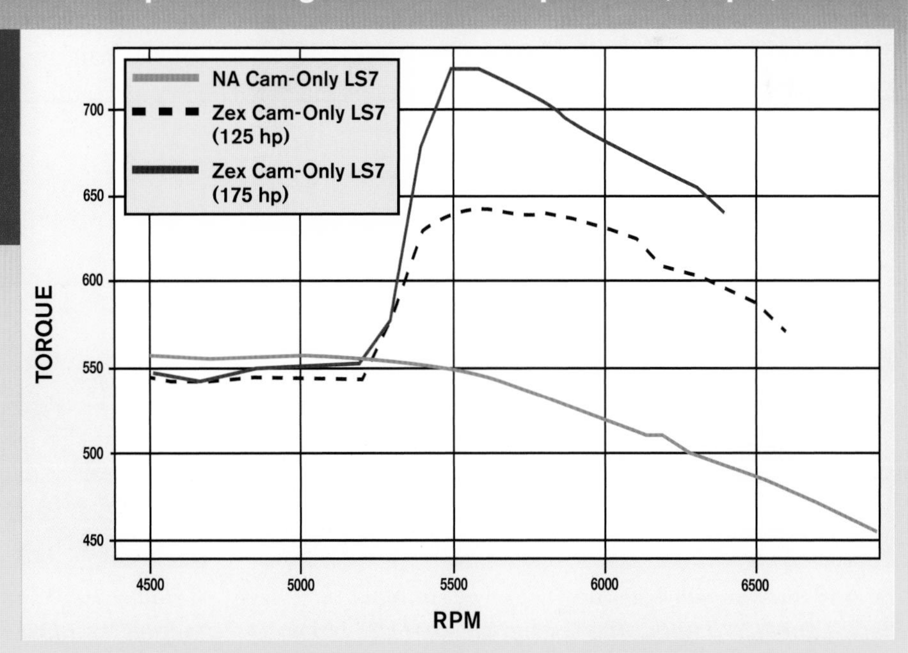

Equipped with the mild Comp cam, the 427 LS7 offered an impressive torque curve, but that curve picked up dramatically after I injected the nitrous. Successive steps in additional torque were offered by 125- and 175-hp jetting. All I did to achieve these power levels was change the jets. No wonder nitrous is so popular among racers, but don't get greedy.

Test 8: 468 LS7 Stroker: NA vs NOS Nitrous Using a 250-hp Shot

Sometimes you just have to go big, and for this test going big meant both the engine and the amount of nitrous. A good rule is that the amount of nitrous you can add is a percentage of the original power output of the engine. The lower the percentage, the easier it is on the engine. Attempting to add 250 hp worth of nitrous to a 250-hp engine (100 percent) would be difficult, if not impossible because the mild engine might not be able to process that amount of nitrous. By contrast, adding that same amount of nitrous to an engine that already exceeds 750 hp means that you drop the percentage to a more manageable 33.

To determine if it was possible for a 750-hp combination to process an additional 33 percent power, I built just such a test engine. Starting with an aluminum LS6 block, the engine was treated to Darton sleeves, which allowed me to push the displacement to 468 ci, thanks to a stroker crank and large bore size. The stroker featured internal components from Lunati, CP, and Total Seal; induction chores were handled by Mast Black-Label heads and a matching single-plane (4500) intake manifold.

Even in NA trim, the LS7-based stroker was no slouch, thanks to a healthy Comp hydraulic-roller cam, a high-flow induction system, and a Holley 1050 Dominator carburetor. Run with an MSD ignition system, American Racing headers, and a Milodon oiling system, the carbureted 468 stroker produced 768 hp at 6,800 rpm and 658 ft-lbs of torque at 5,400 rpm. Torque production hovered near 650 ft-lbs for a solid 1,000 rpm and exceeded 600 ft-lbs from 4,250 rpm to 6,700 rpm.

Obviously, I had a good start to making power, but things became serious once I added the extra 250 hp from the NOS carbureted (cheater) plate kit. With adjustable jetting, I set up the system to add a solid 250 hp, which I hoped would push the stroker over the 1,000-hp mark. After activation, the power needle jumped to 1,006 hp and pushed the torque output to 951 ft-lbs. Peak power exceeded 1,060 hp at the spike, but I selected a number over 1,000 hp after the power curve settled in. Four-digit LS engines don't grow on trees, but the right combination of displacement and nitrous can make serious power.

The 468 stroker was equipped with a set of Mast Black-Label LS7 heads. The Darton-sleeved LS6 block also featured a Lunati crank and rods combined with CP Pistons.

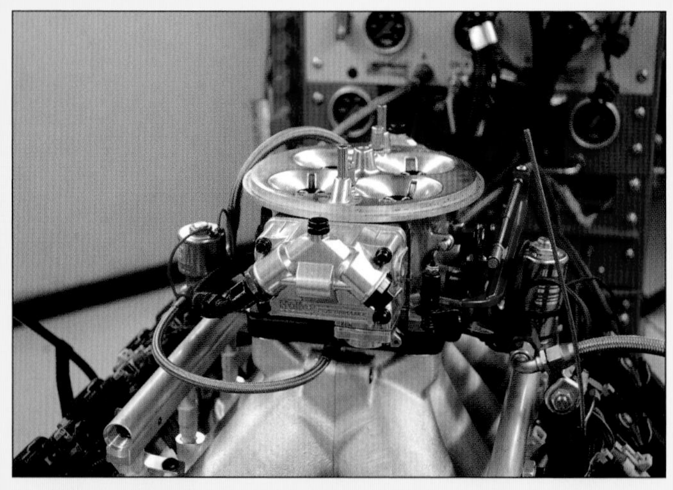

A Holley 1050 Dominator and NOS plate nitrous system topped the Mast two-piece single-plane intake. The Mast intake also featured provisions for injectors, but these were simply plugged during the test.

468 LS7 Stroker: NA vs NOS Nitrous Using a 250-hp Shot (Horsepower)

NA 468 LS7 Stroker: 768 hp @ 6,800 rpm
NOS 468 LS7 Stroker (250-hp shot): 1,006 hp @ 6,200 rpm
Largest Gain: 256 hp @ 6,300 rpm

Not counting the big gain after activation, the 250-hp shot added almost exactly that. Had I not run out of nitrous during the run, the gains would have been consistent to 7,000 rpm. Regardless, the NOS kit added some serious power to an already impressive LS7 stroker combination. Initially in the run, the NOS-equipped LS7 spiked to well over 1,000 hp and then settled to 1,000 hp. Any engine that cracks the 1,000-hp mark is a serious player indeed.

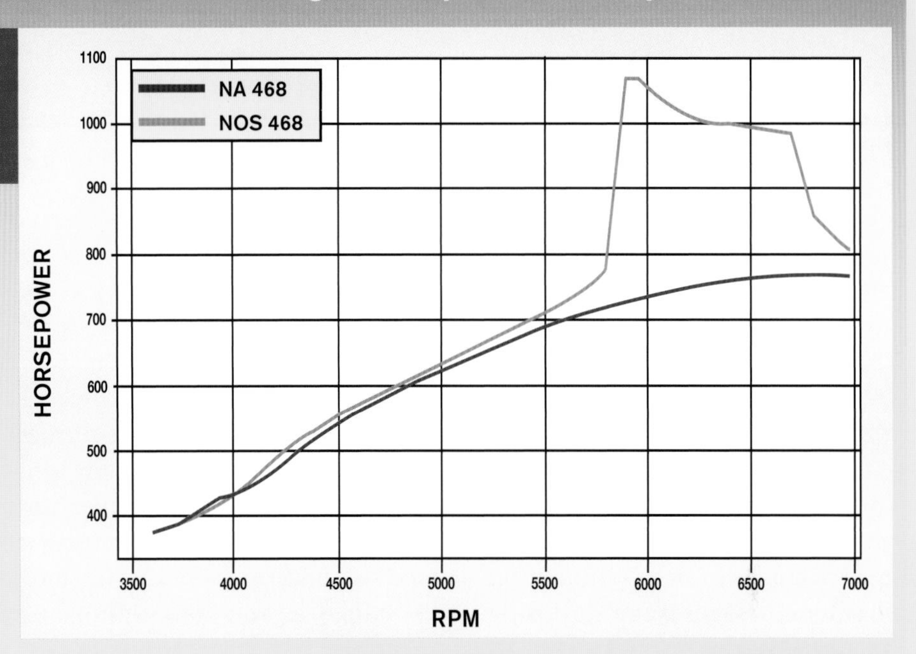

468 LS7 Stroker: NA vs NOS Nitrous Using a 250-hp Shot (Torque)

NA 468 LS7 Stroker: 658 ft-lbs @ 5,400 rpm
NOS 468 LS7 Stroker (250-hp shot): 951 ft-lbs @ 5,900 rpm
Largest Gain: 300 ft-lbs 5,900 rpm

As I have come to expect, I had a huge jump in torque after the initial activation, but because I activated it after 5,252 rpm, the torque gains diminished with engine speed. The fall off at the top of the rev range was the bottle running low on nitrous. Even on a serious 468 LS7 stroker, you can feel an extra 300 ft-lbs of torque!

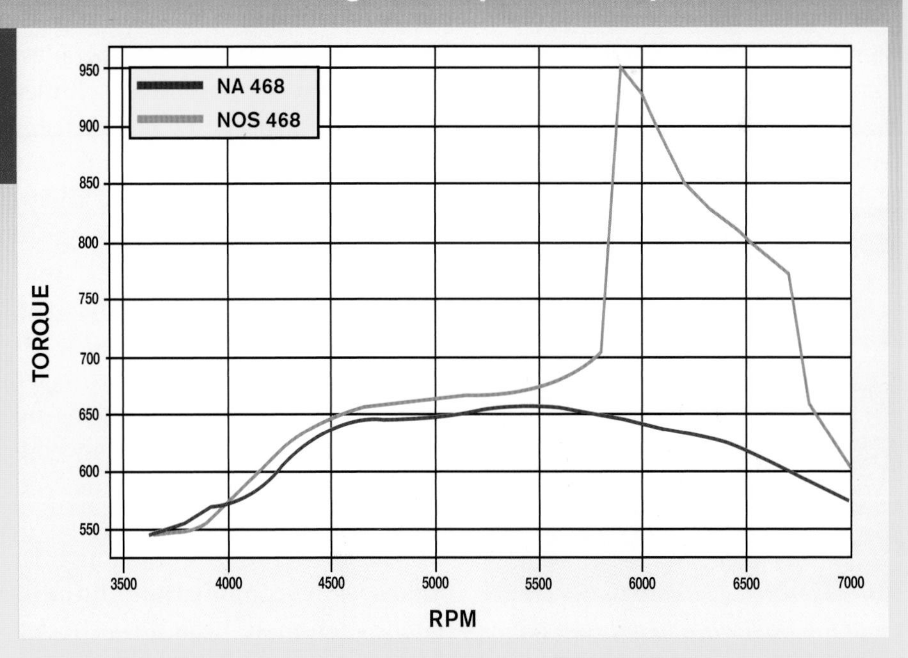

ENGINE BUILDS

This book was purposely divided into individual chapters dealing with all of the performance subsystems, but the truth is that maximum performance comes from a combination of components. This is not surprising because all of the components in the engine, such as head flow, cam timing, and intake (and exhaust) design must all be matched or tuned to optimize power over a given RPM range.

The installation of cam profiles designed to produce high-RPM power on an engine equipped with a stock intake and exhaust system will not likely yield the desired results. The same can be said for a short-runner intake or a header with short primary lengths. The intake runner and exhaust primary lengths must be tuned to optimize power in the same RPM range as the cam timing. Of course, the head flow must support this power and RPM level, though this is rarely a problem with most LS3 or LS7 applications. Everything must work together for maximum power production.

One of the best ways to improve the power output of your LS3 or LS7 is to make it bigger. Adding a stroker crank improves power production through the entire rev range.

The major performance upgrades, such as to intake manifolds, cylinder heads, and (especially) revised cam profiles, offer impressive gains. But as you saw in Chapter 1 and Chapter 2, the gains can be even more impressive on more powerful combinations. The problem is that the stock heads and intake are so good (especially the LS3) that aftermarket versions often don't show their true potential because the stock items are already capable of supporting the current power level.

The answer to this dilemma is to build a wilder combination. Unfortunately, even more head flow, shorter intake runners, and wilder cam timing tend to push power production higher in the rev range and reduce both low-speed power and drivability. Such is the trade-off inherent in power production at any given displacement. At some point, you start rocking the power curve, where more power at one end of the RPM scale comes at the expense of the opposite end. The one way to combat this is to increase displacement.

The benefit of a stroker engine is, of course, the sizable jump in power thanks to the increase in displacement. Let's take a look at a simple example. Suppose you build a 376-inch LS3 producing 495 hp. This equates to 1.316 hp per cubic inch, which are impressive numbers for a stock engine. By the way, the stock LS7 produces 574 hp from 427 ci, putting it slightly ahead with a specific output of 1.344 hp per cubic inch. Calculated using the factory ratings of 430 hp and 505 hp, you get 1.143 for the LS3 and 1.182 for the LS7. Regardless of which calculation you use, if you maintain the same efficiency and increase the displacement, the power output goes up. If you increase the 376-inch LS3 to 416 inches (the most common stroker size), the 495-hp output jumps to 547 hp, all from displacement (assuming you maintain the same level of efficiency). It is obviously possible to increase or decrease efficiency of the larger combination because the bigger engine requires more airflow and cam timing to maintain the specific output. Luckily for LS3 enthusiasts, the stock heads are more than up to the task.

The benefits of increased displacement come not just at the specified horsepower or torque peaks, but through the entire rev range. This extra torque greatly improves part-throttle acceleration, drivability, and can eliminate the need to downshift when passing. In short, the extra power offered by the stroker adds fun. This is especially true when you combine the extra cubic inches with additional airflow, wilder cam timing, and changes to the induction system.

As good as the factory or even modified versions of the LS3 and LS7 are, they get even better when you make them bigger. In fact, a case can be made that the large port volumes and valve sizes of both the LS3 and LS7 heads are better suited to increased displacement. Testing ported heads on an otherwise stock LS3 resulted in very little power gains, but performing the same test on a stroker version resulted in significant gains. Need more proof? How about the fact that I was able to produce nearly 700 hp with a set of stock LS3 heads on a 468 stroker. Will they produce that type of specific output on a 376? Not likely.

To illustrate the benefits and possibilities offered by stroker combinations, I included some LS3- and LS7-based strokers for testing. You might think the 427 LS7 was already technically a stroker and you would be right, but if 427 inches are good, then 468 or (better yet) 495 inches must be even better, right? Even at 427 ci, the LS7 is probably undersized for the factory cylinder head flow, to say nothing of the massive flow potential of some of the aftermarket versions.

A number of the ported LS7 heads reached or exceeded 400 cfm. That was enough to support more than 800 hp on the 495 stroker, but tested on a stock-displacement 427, the cubes just aren't capable of taking full advantage of the available head flow. That is why I used the 495 to test the merits of several different LS7 heads and even stepped up to the 468 stroker to run the LS3 heads (see Chapter 2).

If you combine the added airflow with revised cam timing to the additional displacement, the results are usually a winner. This chapter is full of common (and not-so-common) stroker applications as well as two of the most popular GM crate engines.

Summary

Each successive jump in displacement brought an increase in peak torque production and torque through the entire rev range. Due to the combination of the truck intake, cathedral-port heads, and mild cam timing, the gains offered by the LS3 over the 6.0 truck engine were minimal below 3,000 rpm; but just rev the engine a little and the 6.2 really takes off.

For extra torque, it's hard to beat the extra displacement, cam timing, and compression offered by an LS3.

Displacement is a function of both bore and stroke. To go beyond the 4.030-inch bore of a 6.0 iron block, the 4.070-inch bore of an LS3 block or 4.125-inch bore of an LS7, it becomes necessary to sleeve the block. I employed Darton sleeves on a number of the strokers in this chapter.

As good as a modified 376 LS3 is, it can be even better if you up the displacement to 416 ci. More than just gains in peak power, the stroker kit adds power through the entire rev range.

For maximum displacement, I chose to start the stroker build with this tall-deck, RHS aluminum block. Combining a 4.185-inch bore with a 4.50-inch stroke resulted in an amazing 495 inches.

Test 1: LS3 Chevy Performance Crate Engine

You might be asking why I would highlight something as ordinary as an LS3 crate engine in a chapter on engine buildups, but the answer is simple. As offered by GM Performance (and supplied by my good friends at Gandrud Chevrolet), the LS3 crate engine has a lot going for it. Having run plenty of testing, I know one thing for certain: All crate engines are not created equal. A performance combination that has been remachined, remanufactured, or otherwise reconditioned is simply not the same as a factory-fresh crate engine. There is just something about the word "new" when it's applied to crate engine.

Famous for its lineup of crate engines, the LS3 (PN 19201992) from Chevrolet Performance was nothing less than a brand-new Corvette LS3 plucked from the assembly line. Factory rated at 430 hp and 424 ft-lbs of torque, the LS3 crate engine was much more than just a power number. How many crate engine builders can brag about having literally millions of dollars in research and development of their combinations? Imagine the cost associated with offering an engine on par with a factory LS3. Remember, this is an engine that Corvette and Camaro owners fully expect to run flawlessly for 100,000 miles or more. You just don't find that level of engineering in any other crate engine that didn't originate at the OEM level.

As I mentioned in the introduction of this book, the LS3 has a lot going for it. So much so, that the stock LS3 has made it difficult for the aftermarket to improve upon it. That must really make the designers happy, especially given the many constraints they were up against when it was originally conceived. I know that (as with all LS applications) cam swaps work well on the LS3. The reason they work so well is that the LS3 was factory blessed with amazing head flow and one of the best intake designs in the family. The LS7 takes top honors for factory head flow, but relatively speaking, the LS3 intake is significantly better than the LS7. Both the FAST and MSD intakes offer serious power gains on the LS7, but it is *very* difficult to get extra power without some sort of trade-off on an LS3.

The factory LS3 heads are a similar situation, but not quite as difficult as the intake. The stock heads flow enough to support well over 600 hp, so you better have a serious combination if you plan to run ported versions that flow enough to support more than 700 hp. To illustrate how impressive the factory LS3 was, I compared it (on the graphs) to three other factory offerings: 4.8, 5.3, and 6.0.

One potential weakness of the LS3 (as with all LS engines) was the factory valvesprings. Adequate for stock or low-lift applications, the stock springs should be upgraded when swapping cams.

Even in stock trim, the LS3 crate engine was one impressive piece of factory hardware. When testing, I always replaced the stock drive-by-wire throttle body with a manual version.

LS3 Chevy Performance Crate Engine (Horsepower)

4.8 LR4: 333 hp @ 5,400 rpm
5.3 LM7: 359 hp @ 5,100 rpm
6.0 LQ4: 406 hp @ 5,200 rpm
6.2 LS3: 496 hp @ 5,900 rpm
Largest Gain: 187 hp @ 6,400 rpm

Because the other three factory engines were designed for low-RPM truck applications, it is not surprising that the LS3 offered nearly 100 extra hp over the similar-size LQ4 6.0 and 160 hp over the smallest 4.8. The LS3 heads flowed nearly 70 cfm better than the 317 (or 706 4.8) heads used on the truck engine. When combined with increased displacement, compression, and wilder cam timing, the result was one impressive crate engine.

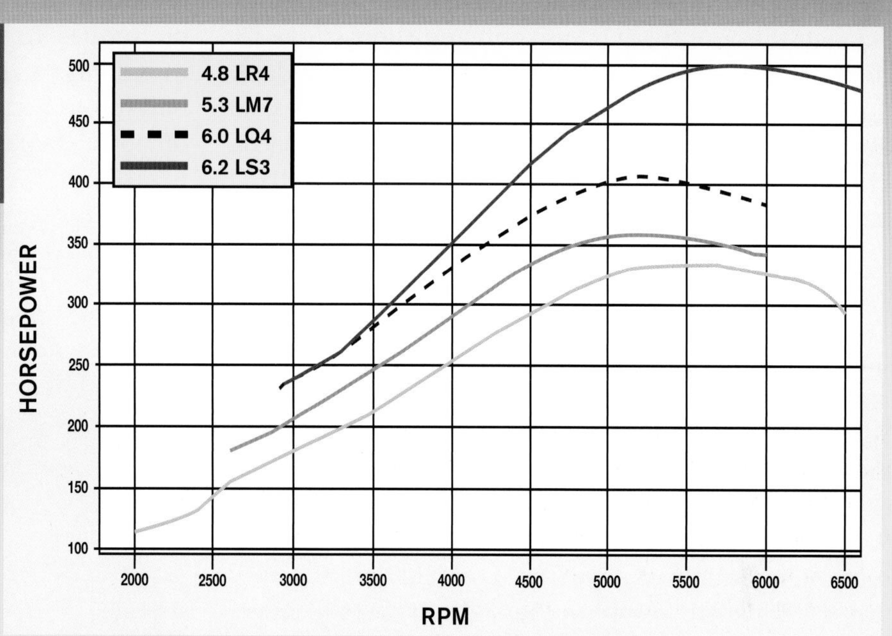

LS3 Chevy Performance Crate Engine (Torque)

4.8 LR4: 343 ft-lbs @ 4,700 rpm
5.3 LM7: 389 ft-lbs @ 4,500 rpm
6.0 LQ4: 439 ft-lbs @ 4,300 rpm
6.2 LS3: 491 ft-lbs @ 4.700 rpm
Largest Gain: 151 ft-lbs @ 4,700 rpm

Each successive jump up in displacement brought with it an increase in peak torque production as well as torque through the entire rev range. Due to the combination of the truck intake, cathedral-port heads and mild cam timing, the gains offered by the LS3 over the 6.0 truck engine is minimal (below 3,000 rpm), but just rev the engine a little and the 6.2 really takes off. For extra torque, it's hard to beat the extra displacement, cam timing, and compression offered by the LS3.

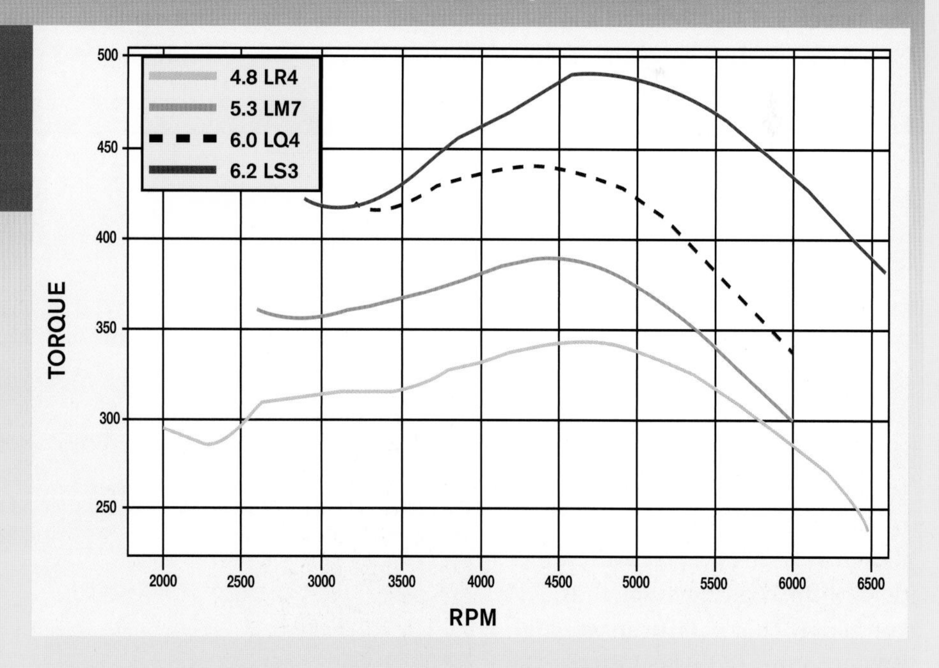

Test 2: 600-hp Short-Stroke LS3

This chapter is chock-full of super strokers, some measuring nearly 500 ci, so I decided that little LS engines (though they are every bit as cool) and short-stroke LS combinations should get some love as well. Why reduce the stroke, you ask? When you shorten the stroke, you decrease the piston speed at any given engine speed. This reduction in piston speed allows for commensurate increases in engine speed.

The real key to the success of a short-stroke application is to combine it with an increase in bore size. Whether you call it a big-bore 4.8 or a short-stroke LS3, the result is a high-RPM engine that is assured plenty of airflow (heads flow better with increased bore size). Small bores limit head flow by shrouding the valves, so most race engine builders look to maximize bore size then reduce stroke for RPM potential. Given the abundance of aftermarket LS blocks and custom cranks, it is possible to build just about any combination. For this test, I decided to combine the shortest factory crank with (nearly) the largest-bore block.

In terms of LS applications, only the 4.8 LR4 and 7.0 LS7 offered different stroke cranks. The 3.622-inch stroke was used in everything else, including the 5.3, 5.7, 6.0, and 6.2 (including the supercharged versions). The larger 7.0 LS7 offered a 4.0-inch stroke, while the 4.8 truck engine dropped displacement courtesy of 3.267-inch stroke. For this buildup, I combined the factory 4.8 crank with a set of 6.30-inch Lunati rods and JE forged pistons.

After adding a custom Comp cam that provided RPM potential, I added TFS Gen X 255 LS3 heads, a Holley Hi-Ram intake, and a pair of 650 XP carburetors. The combination of cam timing, head flow, and induction tuning from the short-runner tunnel-ram intake ensured power production higher in the rev range. After adding a Moroso oiling system, an ATI dampener that's critical at high RPM, and limited travel lifters, the short-stroke LS3 was ready to roll.

After dialing in the combo using an MSD ignition controller (and carb jets), I was rewarded with peak numbers of 614 hp and 446 ft-lbs of torque. To illustrate the trade-offs inherent in a high-RPM small-displacement engine, I compared this buildup to a stock LS3. The bigger stock displacement obviously makes a lot more low-speed torque, but nothing sounds as good a high-RPM LS at full song.

Out came the factory 3.622-inch LS3 crank and in went a 3.267-inch 4.8 crank.

The short stroke was combined with a set of custom forged (6.30-inch) rods from Lunati and JE pistons.

600-hp Short-Stroke LS3 (Horsepower)

Stock LS3: 495 hp @ 5,900 rpm
Short-Stroke LS3: 614 hp @ 7,700 rpm
Largest Gain: 100+ hp @ Above 7,500 rpm

The short-stroke engine ran cleanly up to 8,000 rpm, even with a hydraulic roller cam. When I combined the short stroke with TFS GenX 255 LS3 heads, a custom Comp cam, and a Holley Hi-Ram intake, the little LS3 produced 614 hp at 7,500 rpm. The short stroke helped, but the key to high-RPM power production was really the cam, heads, and intake.

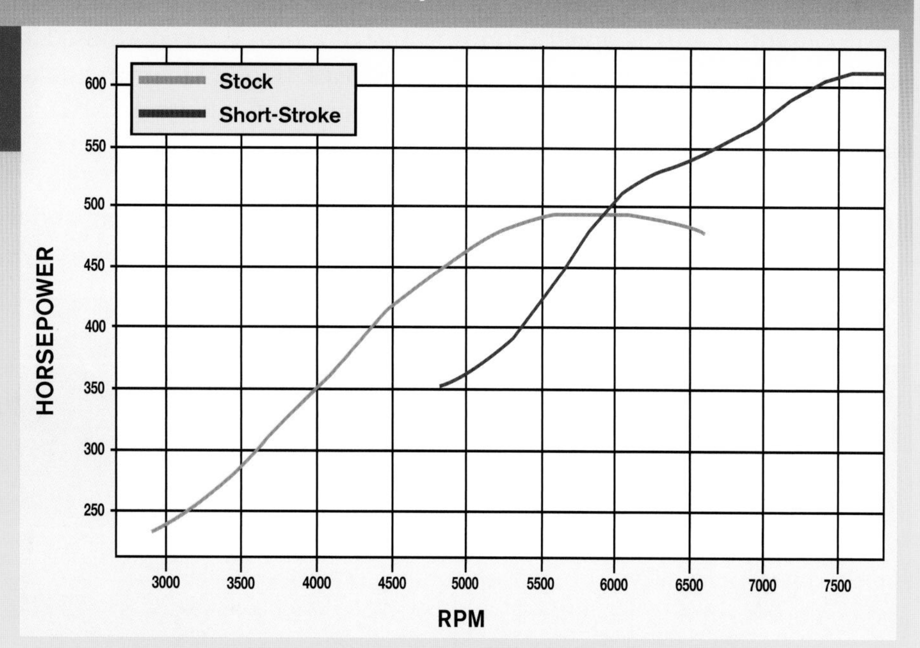

600-hp Short-Stroke LS3 (Torque)

Stock LS3: 489 ft-lbs @ 4,700 rpm
Short Stroke LS3: 446 ft-lbs @ 6,100 rpm
Largest Gain: (-100) ft-lbs @ Below 5,000 rpm

Short-stroke engines equipped with rectangular-port heads, tunnel-ram intakes, and wild cam timing do not excel at torque production. Despite exceeding 600 hp, the short-stroke LS3 only managed to produce 446 ft-lbs of torque (40 less than a stock LS3). Had I run the short-stroke engine below 4,800 rpm, it would be lower by 100 ft-lbs. If you want that high-RPM scream, it is going to cost you torque.

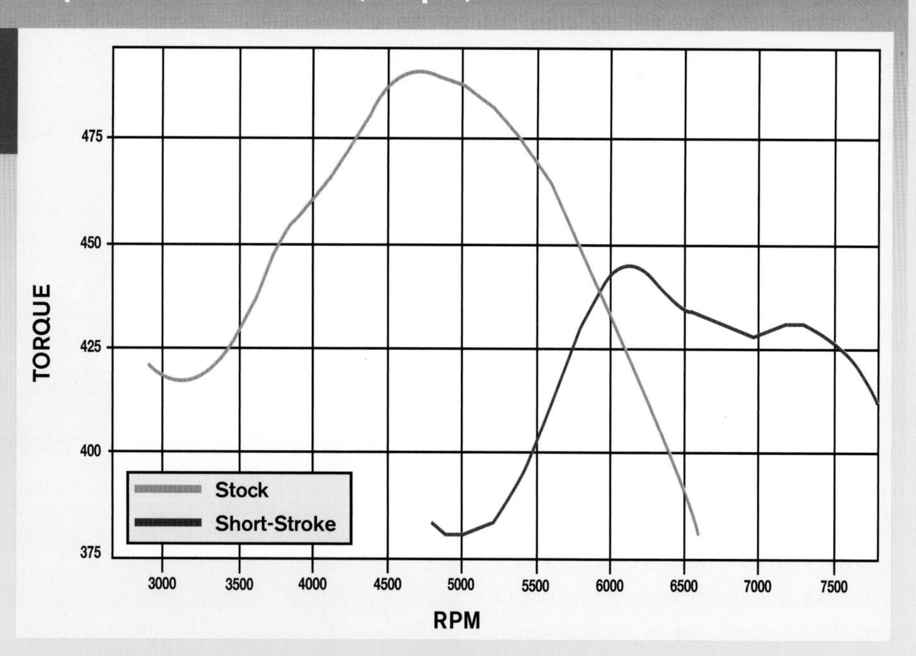

Test 3: Stock LS3 vs 416 LS3 Stroker

I know that for the testing in this book (and every LS article ever written), the best way to improve the power output of an LS3 application is to install a performance camshaft. A cam swap alone can increase the power output of an LS3 by 50 to 60 hp or more, but what happens when you want even more? More than just more, what happens when you want to improve the power output without sacrificing drivability?

Sure, more power can be had from an LS3 with even wilder cam timing, but this would also require the installation of a set of aftermarket, flat-top pistons with valve reliefs to allow for the increased duration (piston-to-valve clearance is the limiting factor in terms of cam timing on a stock LS3). If you are going to install pistons, you might as well throw in forged rods as well. Even then, you are going to start trading drivability for increased peak power. If you are thinking about replacing the pistons anyway, just step up to an LS3 stroker and be done with it.

The great thing about a stroker LS3, with the 416 being the most common displacement, is that the extra power can come with no penalty in drivability. The extra cubes actually add power and increase drivability because they tend to tame any cam profile used on the stock displacement or allow wilder cam timing for even more power. The extra power offered by the displacement change also allows changes in static compression ratio, which can be used to implement forced induction or nitrous.

The 416 stroker used for this test combined a Scat 4.0-inch crank and 6.125-inch rods with a set of 4.070-inch JE Asymmetrical forged pistons. The dished pistons actually reduced the static compression ratio slightly, but the increase in displacement more than made up for the change. Designed to be a street engine, I installed a Comp hydraulic roller cam that featured a .614/.621-inch lift split, 231/239-degree duration split, and 113-degree LSA using Comp hydraulic roller lifters. Headgear included CNC-ported L92 heads from GM Performance using Fel Pro MLS head gaskets and ARP head studs. Finishing touches included an Edelbrock Victor Jr. intake and a Holley 950 XP carburetor, along with an MSD ignition controller.

Run with a set of 1⅞-inch American Racing headers, the 416 stroker produced peak numbers of 628 hp at 6.700 rpm and 544 ft-lbs of torque at 5,300 rpm. Best of all, it was ready if I decided to apply boost (I did).

The stroker was built using a set of 4.070-inch JE Asymmetrical forged pistons. The pistons were combined with a forged 4.0-inch stroker crank and rods from Scat.

The stroker was topped with a set of CNC-ported L92 heads from GM Performance (supplied by Gandrud Chevrolet).

Stock LS3 vs 416 LS3 Stroker (Horsepower)

Stock LS3: 495 hp @ 5,900 rpm
416 LS3 Stroker: 628 hp @ 6,700 rpm
Largest Gain: 149 hp @ 6,600 rpm

Compared to a stock LS3, the stroker offered an extra 35 to 40 hp up to 5,000 rpm, but the gains increased dramatically up to 6,700 rpm. The combination of displacement, head flow, and cam timing really showed itself because the 416 stroker produced 628 hp. Measured peak to peak, this was a gain of 136 hp, but the gains were even higher past the power peak.

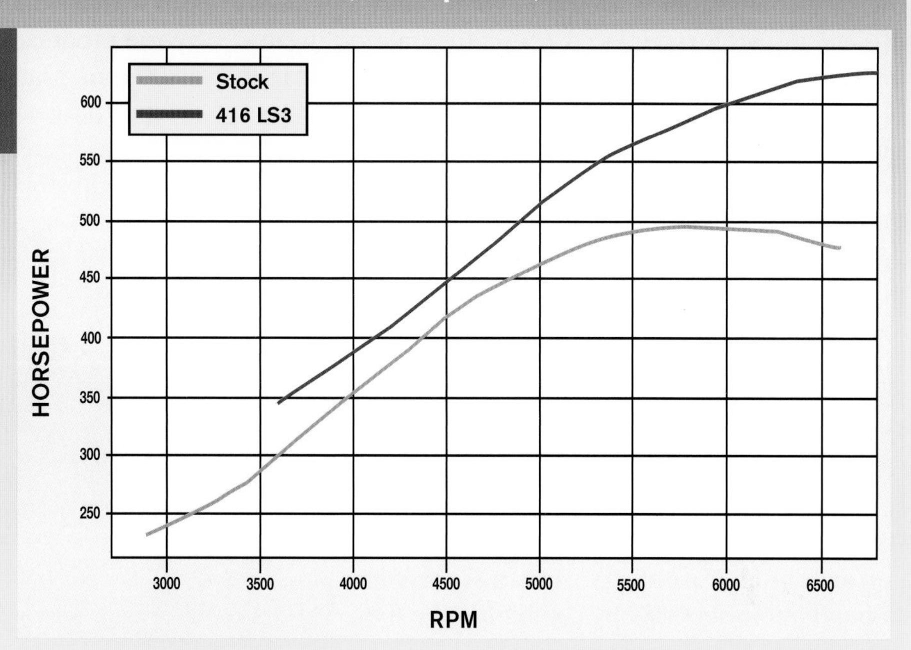

Stock LS3 vs 416 LS3 Stroker (Torque)

Stock LS3: 489 ft-lbs @ 4,700 rpm
416 LS3 Stroker: 544 ft-lbs @ 5,300 rpm
Largest Gain: 84 ft-lbs @ 5,800 rpm

Bigger engines make torque; it's a fact of life, but there is more to the torque gains than simple displacement on this 416 stroker. The extra cubes were combined with increased compression, better head flow, and wilder cam timing. The use of a single-plane Victor Jr. intake actually hurt torque production (see Chapter 1) compared to a long-runner EFI intake, but the 416 stroker still offered more torque than the stock LS3. The 416 produced 544 ft-lbs, compared to 489 ft-lbs for the stock LS3.

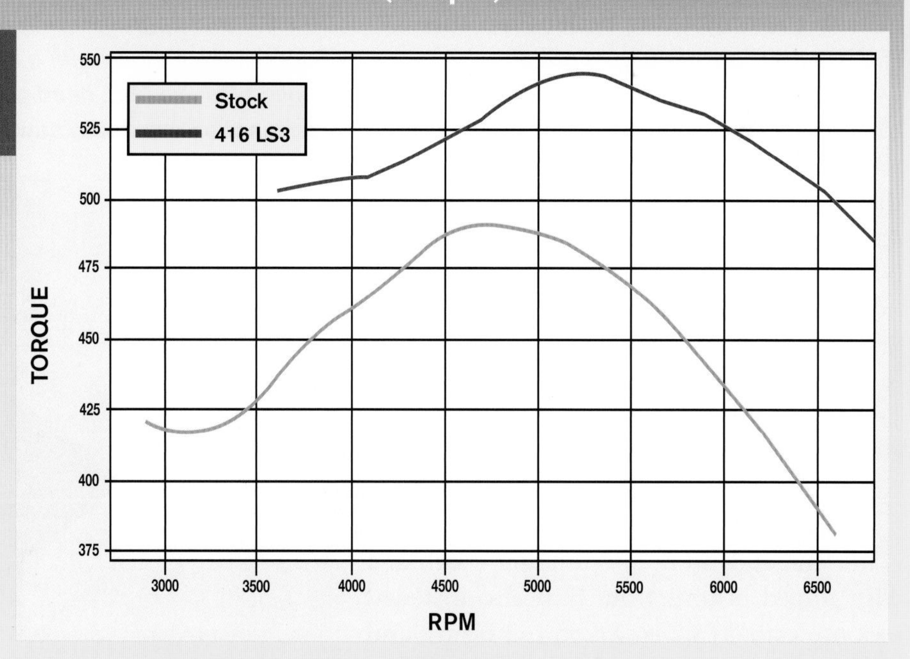

Test 4: LS3 vs 408 LS3 Hybrid Stroker

One of the reasons I love the LS engine family is the interchangeability. Despite evolutionary upgrades as General Motors replaced the cathedral-port heads on performance applications with rectangular-port LS3 heads, the important parts all interchange. One of the popular "family" upgrades is to upgrade cathedral-port heads with LS3 heads. Why go to the trouble, you ask? The best cathedral-port heads flow less than 250 cfm, while the LS3 heads flow a whopping 315–320 cfm.

What LS engine couldn't use an extra 70 cfm per cylinder? While this seems like a no-brainer, there is a catch because the big-valve, LS3 heads were designed for the 4.065-inch bore. Although they can't be installed on the small-bore 4.8 or 5.7 applications, they fit on the larger 4.0-inch bores used on the iron truck and aluminum LS32 blocks. Thus, it is possible to build a low-buck LS3 by combining the 6.0 iron block with the top-end of the LS3.

Of course, the slightly smaller 6.0 produces less power than the 6.2 LS3, but the combination is considerably less expensive than purchasing a used or crate LS3 outright. In fact, the only thing better than this 6.0/LS3 hybrid is a stroker version!

Building a stroker version of the 6.0 hybrid is simple. The 6.0 iron block readily accepts the same 4.0-inch stroker crank, but the bore size must be kept to a maximum of 4.030 instead of the 4.070 used on the LS3 block. Again, the extra displacement offered by the larger-bore aluminum block ultimately makes more power, but the iron-block route is considerably less expensive and stronger. This comes into play if you decided to run a turbo, blower, or heavy dose of nitrous on your stroker (which I did).

This combination included a forged crank and rods from Speedmaster combined with forged pistons from JE. I also installed a Comp cam, TFS Gen X 255 LS3 heads, and a FAST LSXR intake. I selected a dished piston to keep the static compression reasonable for boost, but the combination was plenty healthy even before I added the Vortech supercharger. Run with a Holley HP management system, the 408 LS3 hybrid produced 607 hp at 6,200 rpm and 566 ft-lbs at 4,800 rpm. I later ran this over the 1,000-hp mark with a supercharger. Remember, the more power you start with, the more power you make under boost, and an LS3 hybrid stroker is a great way to make power.

The 6.0 iron block is a perfect candidate for an LS3 head swap. This 6.0 was upgraded with a 4.0-inch stroker crank and rods from Speedmaster, along with a set of JE forged (dished) pistons. I also applied Fel Pro MLS head gaskets and ARP head studs because this combination eventually saw both boost and nitrous.

The stroker was equipped with a Comp cam (PN 54-461-11) that offered a .624-inch lift, 239/247-degree duration split, and 114-degree LSA.

LS3 vs 408 LS3 Hybrid Stroker (Horsepower)

Stock LS3: 495 hp @ 5,900 rpm
408 Stroker Hybrid: 607 hp @ 6,200 rpm
Largest Gain: 117 hp @ 6,400 rpm

Because none of the stock cathedral-port heads flow as much air as the stock LS3 heads, it is common practice to upgrade cathedral-port engines with rectangular-port heads. This is only possible on 6.0 blocks (iron or aluminum) because they feature the necessary 4.0-inch bore required for the larger LS3 valves. I applied a set of TFS Gen X 255 heads to a 4.030 6.0 stroker with excellent results. The LS3 6.0 hybrid combination produced 607 hp, or as much as 117 hp more than a stock LS3.

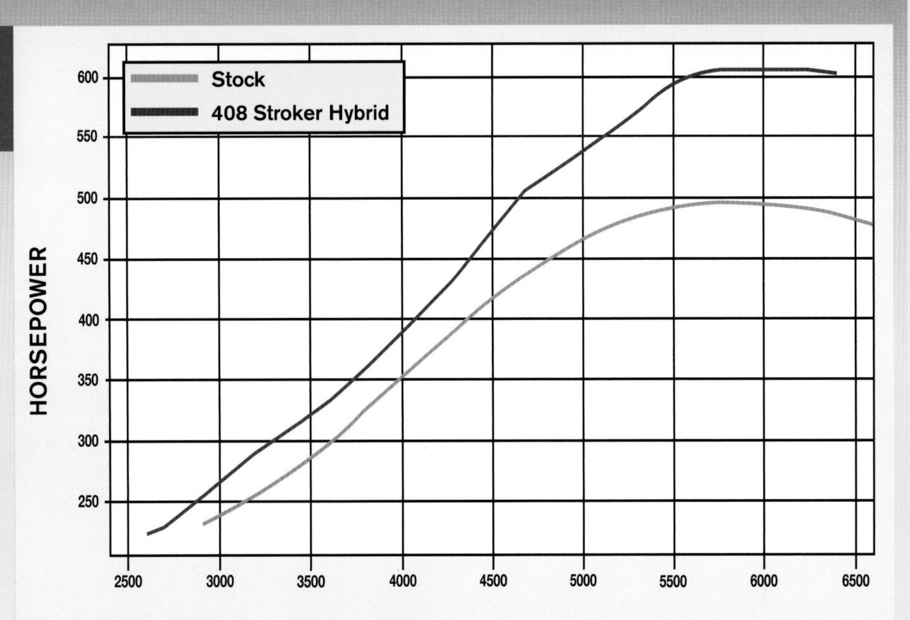

LS3 vs 408 LS3 Hybrid Stroker (Torque)

Stock LS3: 489 ft-lbs @ 4,700 rpm
408 Stroker Hybrid: 566 ft-lbs @ 4,800 rpm
Largest Gain: 97 ft-lbs @ 5,500 rpm

You should expect big peak torque gains when you increase displacement, but the real benefit is the extra torque offered through the entire rev range. It's hard to argue with the extra 50 to 60 ft-lbs offered below 4,500 rpm, to say nothing of the nearly 100 extra ft-lbs past 5,500 rpm. Adding stock or aftermarket LS3 heads to a 6.0 block is an affordable way to produce a solid stroker. Now all it needs is boost.

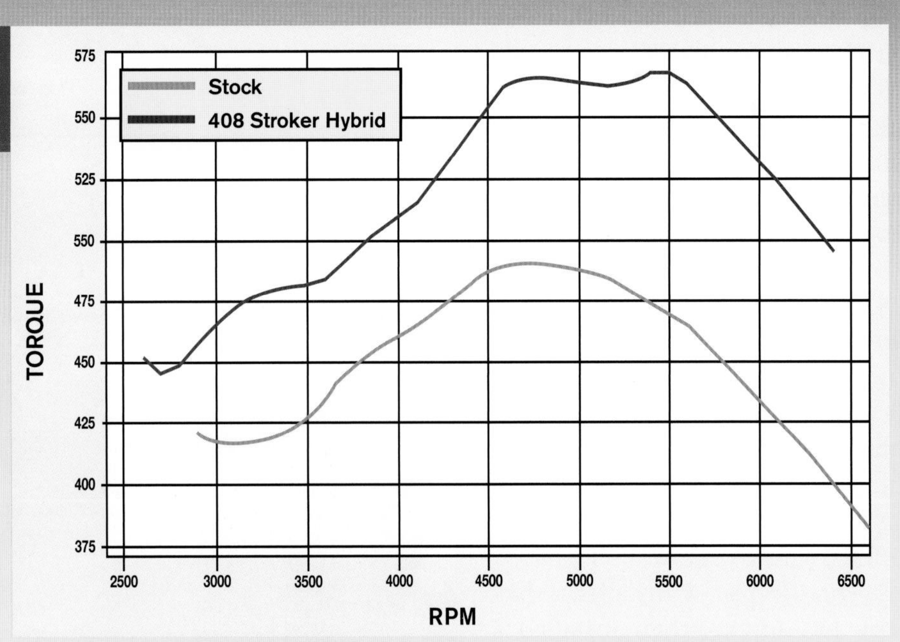

Test 5: Supercharged GM B15 LSX 376

One thing I like about purchasing a crate engine from GM Performance is that you can buy with confidence. Case in point: the B15 LSX crate engine designed specifically for boost. The designation itself stands for Boost 15 or 15 psi of boost, but rest assured the LSX combination can withstand considerably more than 15 psi. How do I know? Well, this particular test was run with more than 22 psi, but before I start adding boost, let's take a closer look at what I added it to.

The B15 LSX crate engine included a number of desirable features, including a six-bolt, iron LSX block for strength. To this, General Motors added a forged crank and low-compression pistons (9.0:1 with the supplied 68-cc heads). The choice of cams was the hottest LS factory cam available, the LS9. Designed for the supercharged ZR1, the cam was a natural for the blower-specific crate engine.

With a solid bottom end, General Motors topped the LSX block with a set of six-bolt aluminum LSX/LS3 heads. The as-cast LSX heads included valvesprings capable of supporting the .560-inch-lift cam, but a spring upgrade is necessary if you perform a cam swap. The heads also received factory LS3 rockers and pushrods, all covered by a set of what might be the nicest covers offered for any LS. Because General Motors figured many enthusiasts would be running blowers on a variety of chassis combinations, the engine was offered sans intake and oil pan.

For the test, I installed a factory LS3 intake, then installed a 4.0 Whipple supercharger. Run with the stock intake, the 9.0:1 376 (basically LS3) produced 478 hp at 6,100 rpm and 446 ft-lbs of torque at 5,100 rpm. After adding the Whipple supercharger, the power output of the supercharged B15 improved to 927 hp and 808 ft-lbs of torque. After the results of this test, maybe they should rename the B15 the B22, B23, or even B24.

If you are thinking about adding boost to your LS, look no further than the GM B15 LSX combination.

The boost supplied to the GM B15 LSX crate engine from the Whipple supercharger was adjustable using changes to the blower pulley.

Supercharged GM B15 LSX 376 (Horsepower)

NA GM B15 LSX: 478 hp @ 6,100 rpm
Supercharged GM B15 LSX: 927 hp @ 6,500 rpm
Largest Gain: 427 hp @ 6,400 rpm

The amazing B15 LSX 376 crate engine from General Motors was built for boost. The combination of an LSX block stuffed with a forged crank and pistons meant it could withstand some serious boost (more than the 15 psi it was named for). Adding more than 22 psi to the B15 from a Whipple supercharger resulted in more than 925 hp.

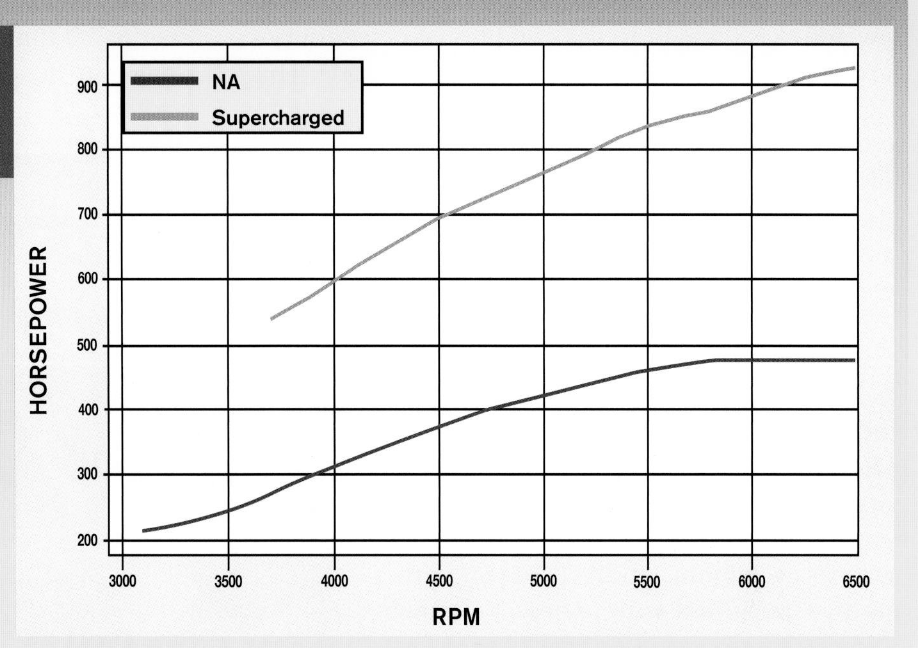

Supercharged GM B15 LSX 376 (Torque)

NA GM B15 LSX: 446 ft-lbs @ 5,100 rpm
Supercharged GM B15 LSX: 808 ft-lbs @ 4,500 rpm
Largest Gain: 366 ft-lbs @ 4,600 rpm

As you saw in Chapter 5, positive displacement superchargers offer immediate boost response and huge torque gains. Of course, the engine must be made to handle the abuse and the GM B15 LSX was designed specifically for boost. Hardly stellar in NA trim, the low-compression B15 was just begging for boost. After adding the Whipple supercharger, the B15 came alive with more than 800 ft-lbs of torque.

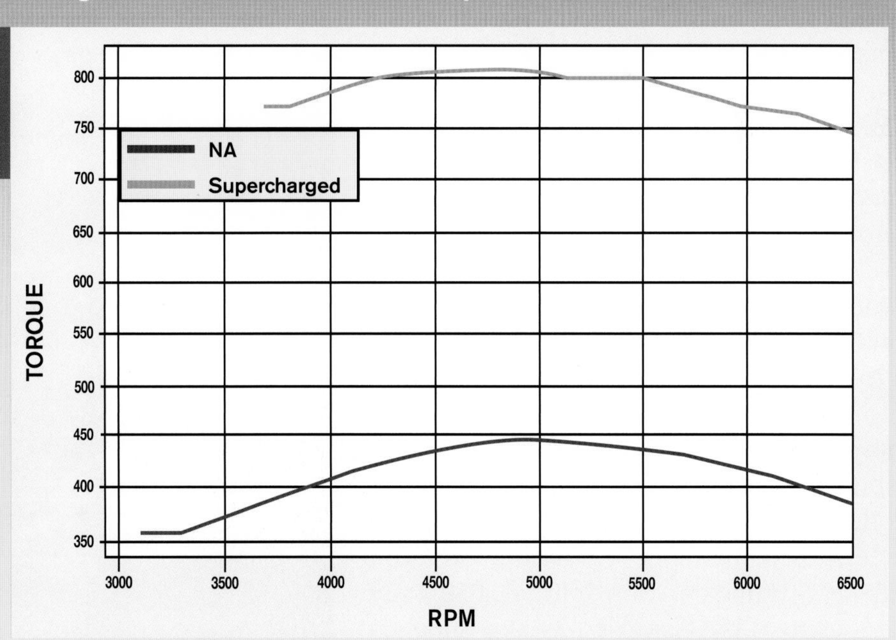

Test 6: Stock vs 468 LS3 Stroker

One of the things you should take away from this book is the fact that stock LS3 heads offer massive head flow and it is difficult to coax additional power out of most combinations with a cylinder head upgrade. This is especially true of stock displacement applications because (on paper) the factory LS3 heads flow enough air to support nearly 650 hp. Truth be told, I produced just under 700 hp using stock LS3 heads on the 495-inch stroker listed in Test 8. The combination of large port volumes, large valve sizes, and plenty of flow means the LS3 heads are best used on large-displacement applications. That is where stroker engines come into play. The increased displacement allows the combination to take full advantage of the elements that the LS3 heads have to offer. This becomes even more critical when you further improve the airflow with ported or aftermarket heads such as the Mast Black-Label heads I employed on this stroker.

To take full advantage of the available airflow, I built a stroker LS by increasing both the bore and stroke of an LS3. I didn't start with a 4.065-inch LS3 block; I started with a smaller-bore LS6 block.

The block was machined to accept Darton sleeves, which allowed me to increase the bore size to a full 4.185 inches. The Darton-sleeved block was combined with a Lunati 4.25-inch stroker crank, forged K1 connecting rods, and Wiseco forged flat-top pistons. The combination resulted in a finished displacement of 468 ci with a static compression ratio of 12.25:1. I topped the stroker not with stock LS3 heads but rather a set of Mast Black-Label heads that flowed just over 380 cfm. Finishing touches included a Comp hydraulic roller (PN 305LRR HR15) cam that featured a .624-inch lift (both intake and exhaust), 255/271-degree duration split (at .050), and 115-degree LSA. Comp Cams also supplied the lifters, double-roller timing chain, and hardened pushrods. The induction system included a Mast CNC-ported, single-plane intake, and Holley 1050 Dominator carburetor.

After break-in and tuning, the stroker produced 761 hp at 6,500 rpm and 645 ft-lbs of torque at 5,100 rpm. I later ran this stroker engine with a healthy shot of nitrous.

The aluminum LS6 block was first treated to Darton sleeves to allow an increase in bore size, then combined with a Lunati stroker crank and forged K1 rods. The bottom end also featured a Milodon pan, pickup, and windage tray.

The stroker required plenty of airflow so I installed these Mast Black-Label LS3 heads. With intake flow that topped 380 cfm, the heads were easily capable of supporting the expected power level.

Stock vs 468 LS3 Stroker (Horsepower)

468 Stroker LS3: 761 hp @ 6,500 rpm
Stock LS3: 495 hp @ 5,900 rpm
Largest Gain: 278 hp @ 6,500 rpm

Increasing the power output of the already impressive LS3 by as much as 278 hp is a simple matter of changing everything. I increased displacement, added better heads, wilder cam timing, and a high-flow induction system. I also increased the static compression ratio and replaced the fuel injection with carburetion. When it was all said and done, the 468 stroker pumped out 761 hp compared to 495 hp for the stock LS3.

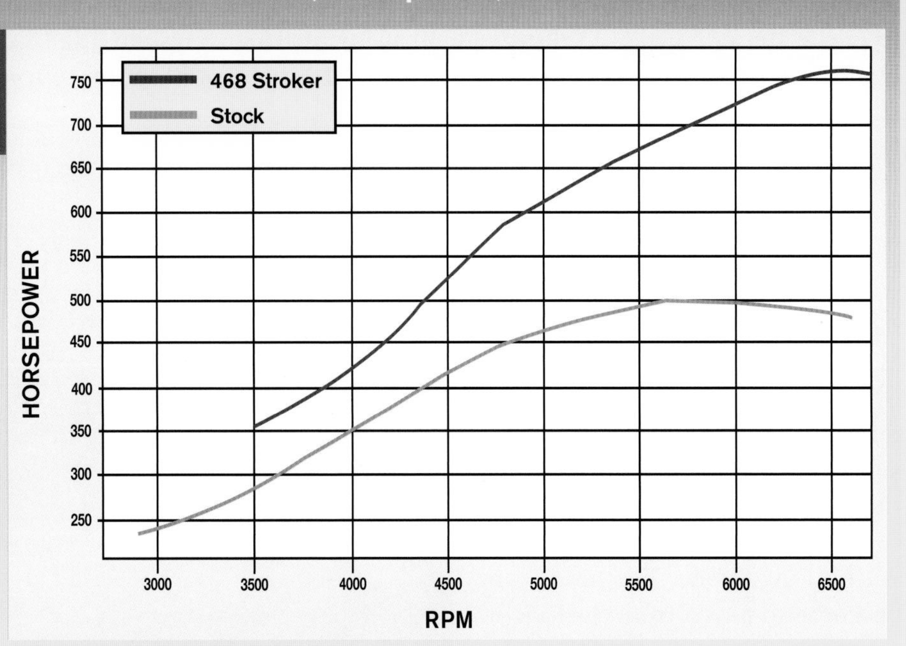

Stock vs 468 LS3 Stroker (Torque)

468 Stroker LS3: 645 ft-lbs @ 5,100 rpm
Stock LS3: 489 ft-lbs @ 4,700 rpm
Largest Gain: 176 ft-lbs 5,600 rpm

Big displacement was only part of the reason for so much increased torque production. The additional head flow, increased compression, and wilder cam timing all combined to increase the specific torque production from 1.30 ft-lbs per cubic inch (on the stock LS3) to 1.40 ft-lbs per cube on the stroker. More displacement always brings more torque, but combining it with increased efficiency pushes production to the next level. The 461 offered 645 ft-lbs compared to 489 ft-lbs for the stock LS3.

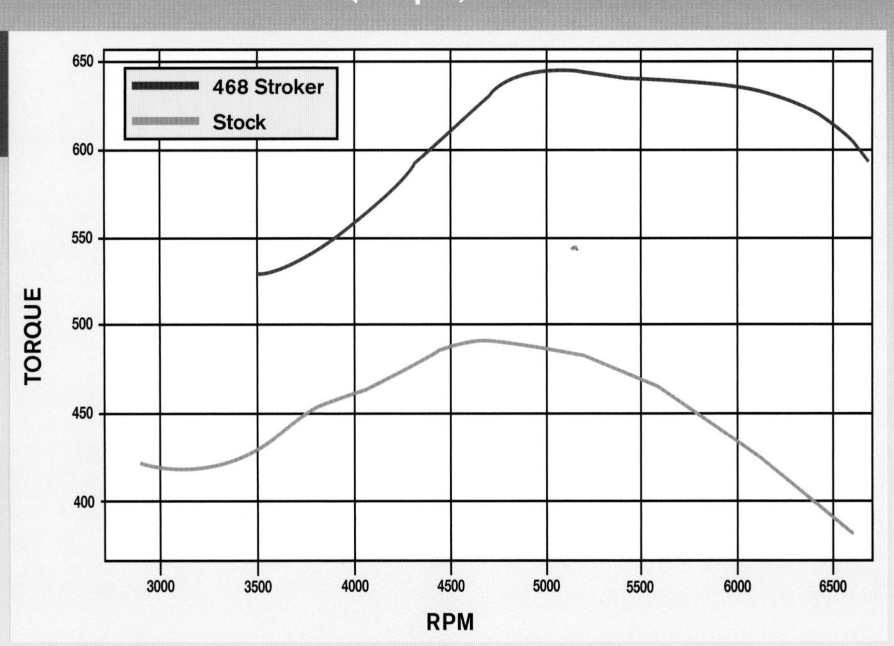

Test 7: LS7-Headed LSX 427

Although LS3s and their stroker variants are commonplace, LS7s are considerably less so. It is possible to create the 427 (or 7.0L) displacement using an LS3 block, by combining a 4.070-inch bore with a 4.10-inch stroke crank, but the true 427 differs in its bore and stroke combination. The factory LS7 starts with a 4.125-inch bore block then adds a 4.0-inch crank. Compared to the long-stroke LS3 combo, the big-bore version offers increased airflow and improved ring control by keeping the piston skirt contained in the sleeve at BDC. The shorter factory cylinders used in the LS3 block allow the piston skirt to come out of the bore at BDC with a long stroke. The result can be a drop in power from the lack of ring seal and airflow, to say nothing of the RPM potential offered by the shorter stroke.

Knowing this, I assembled the 427 using the tired-and-true 4.125x4.0 combination. Starting with an LSX block, I combined a Lunati Voodoo forged crank with Bullet-series CP forged pistons and 6.125-inch Carrillo rods. This combination produced a near bulletproof bottom end capable of supporting all but the wildest power adders I might throw at it. The forged rotating assembly and LSX block were then treated to a trio of power producers, including a Stage IV LS7 cam from BTR, a set of CNC-ported Gen X 260 LS7 heads from TFS, and an Atomic LS7 intake from MSD. Finishing touches on the Magnificent LS7 included an aluminum front cover from Comp Cams, an ATI dampener, and complete Moroso oiling system.

Run on the dyno with a 105-mm Holley throttle, FAST 75-pound injectors, and XFI/XIM management system, the LSX 427 produced 656 hp at 6,400 rpm and 593 ft-lbs of torque at 5,100 rpm.

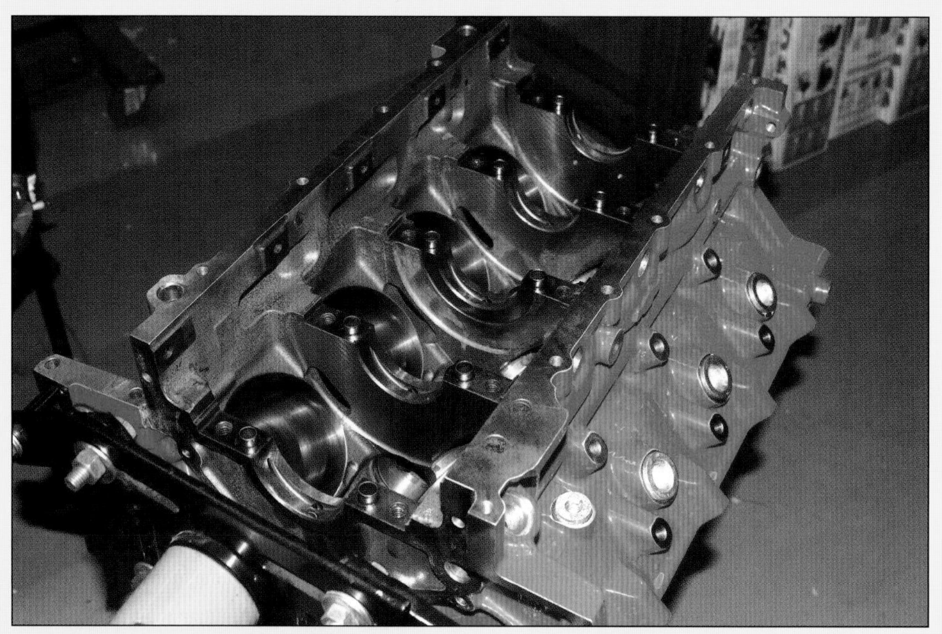

The 427 LSX (code-name Magnificent LS7) started with a solid foundation in the form of an iron LSX block from Gandrud Chevrolet. Capable of supporting four-digit power levels, the LSX block was the perfect choice for this buildup.

The LSX block was treated to a Lunati forged crank, Carrillo rods, and CP pistons, along with ARP head studs and Fel Pro MLS head gaskets.

LS7-Headed LSX 427 (Horsepower)

Stock LS7: 574 hp @ 6,100 rpm
TFS-Headed LSX 427: 656 hp @ 6,400 rpm
Largest Gain: 82 hp @ 6,100 rpm

Compared to the stock LS7, the LSX 427 offered more power everywhere, from 3,500 through 6,500 rpm. Tested at the top of the rev range, the LSX offered nearly 100 extra horsepower, despite the use of a wet-sump oiling system (the factory LS7 runs a dry sump).

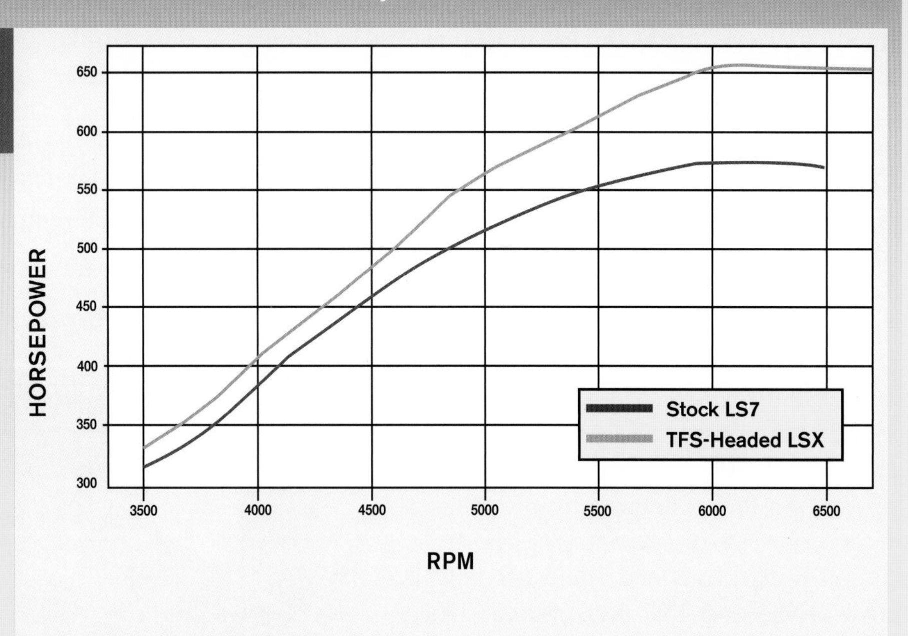

LS7-Headed LSX 427 (Torque)

Stock LS7: 543 ft-lbs @ 4,000 rpm
TFS-Headed LSX 427: 593 ft-lbs @ 5,100 rpm
Largest Gain: 70 ft-lbs 5,900 rpm

Torque gains were significant, with the LSX increasing torque production from 543 to 593 ft-lbs. More important, torque production was up everywhere, which is always a good sign, although the largest gains occurred past 4,500 rpm.

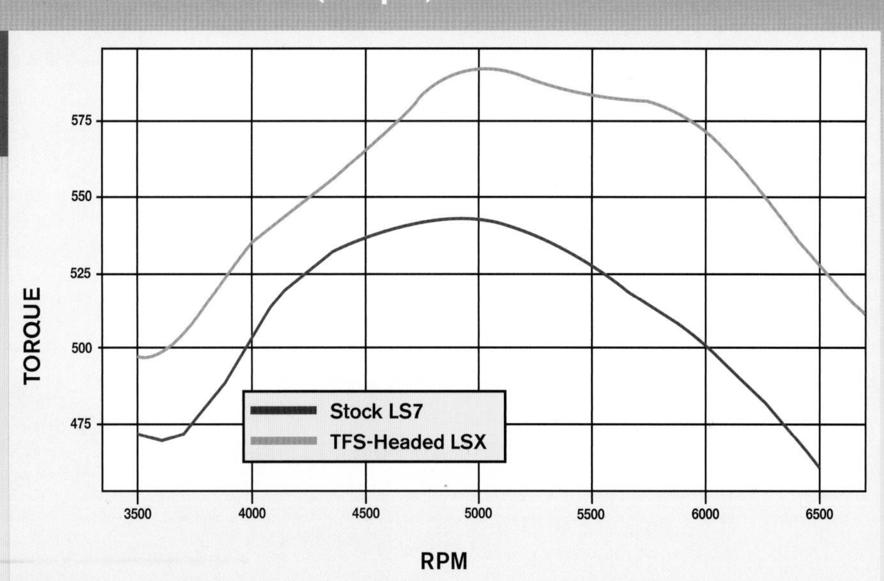

Test 8: RHS 495 LS7 Stroker

I saved the biggest and baddest stroker build for last. The factory LS7 is not only the biggest and baddest small-block ever made, but the most powerful (naturally aspirated) production engine ever. None of the muscle car small-blocks even come within 100 hp (actually more like 200 hp) of the real output of the aluminum 427, and only the limited-production, L88 and ZL1 427 big-blocks come close, but not equipped with their factory cast-iron exhaust manifolds.

Only recently did the NA 427 lose the power output crown to the 6.2 LS9, but it took supercharging to add the extra 132 hp. The factory LS7 combined a 4.125-inch bore with a 4.0-inch forged crank, titanium rods, and coated 11.0:1 pistons. Power production from the 427 came courtesy of factory CNC-ported (raised port) heads and the most powerful production LS7 ever offered. The result was a rated power output of 505 hp and 470 ft-lbs, but tested on the dyno with headers, no accessories (or air intake system) and an optimized tune, it produced 574 hp and 543 ft-lbs of torque.

As much as I liked the biggest and baddest factory LS stroker, I set out to make one bigger and badder. To make it bigger, I started with an aluminum RHS block, but not the standard deck height; I stepped up to the tall-deck version. The tall-deck RHS block allowed me to combine a 4.185-inch bore with a massive 4.50-inch stroke. The result was a finished displacement of 495 ci.

The rotating assembly included a Lunati crank, K1 rods, and Wiseco, 13.5:1 pistons. Since I had a big engine, I selected the largest off-the-shelf hydraulic roller cam available in the Comp Cams catalog. This one (PN 309LRR HR15) offered a .660-inch lift, 259/275-degree duration split, and 115-degree LSA. The cam was combined with a set of Mast Motorsport's Black-Label LS7 heads, a matching (high rise) single-plane intake, and a Holley 1050 Ultra Dominator carb. Unlike the dry sump used on the factory LS7, this stroker relied on a traditional wet sump that featured a pan, pickup, and windage tray from Moroso.

After tuning, the big stroker pumped out some impressive numbers, topping the 800-hp mark with peaks of 810 hp at 6,600 rpm and 726 ft-lbs of torque at 5,300 rpm.

A number of LS7 heads were available from a variety of sources. These heads included both dedicated and ported factory castings. I chose a set of Black-Label LS7 heads from Mast for this 495-inch stroker.

It took some hammer time, handiwork, and washers to finally get the windage tray to fit with the 4.50-inch stroker crank. The 495 featured an RHS tall-deck block stuffed with internal components from Lunati, Wiseco, and K1.

RHS 495 LS7 Stroker (Horsepower)

Stock LS7: 574 hp @ 6,100 rpm
RHS 495 Stroker: 810 hp @ 6,600 rpm
Largest Gain: 234 hp @ 6,400 rpm

It is amazing that the (small-block) LS engine family can be punched out to big-block displacements. Credit the tall-deck RHS aluminum block for the ability to swallow stroker cranks with 4.5 inches (or more) of stroke length. This combined with a possible bore size of 4.20 inches equates to displacements exceeding 500 cubes. This build featured a 4.185-inch bore with a 4.50-inch stroke for a finished displacement of 495 inches. Equipped with Mast LS7 heads, CNC single-plane intake, and healthy Comp cam, the big stroker produced 810 hp.

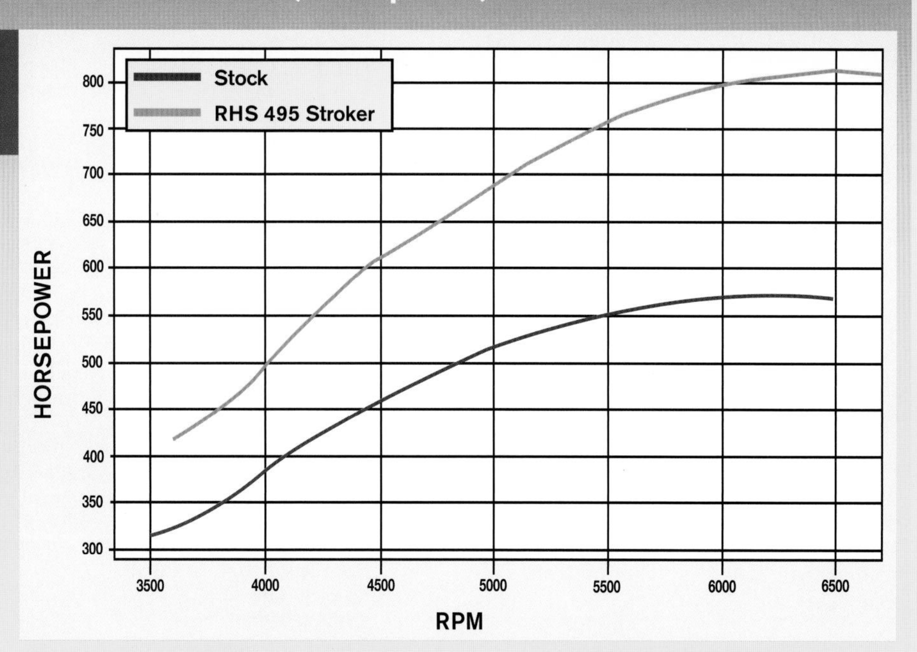

RHS 495 LS7 Stroker (Torque)

Stock LS7: 543 ft-lbs @ 4,900 rpm
RHS 495 Stroker: 726 ft-lbs @ 5,300 rpm
Largest Gain: 192 ft-lbs 5,400 rpm

Compared to the stock 427 LS7, the 495 offered huge torque gains. You know you've made some serious progress when the torque gains exceed 150 ft-lbs. The great thing about stroker engines, especially those sporting big-block dimensions, is that they don't have to run excessive engine speed to make power. Even with a single-plane intake, this stroker thumped out 726 ft-lbs of torque.

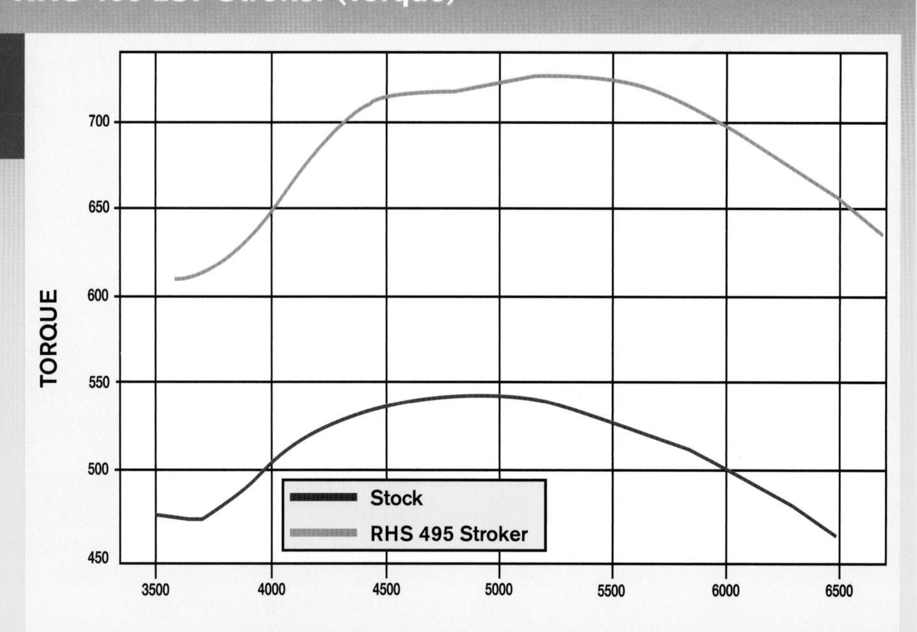

Accufab, Inc.
1326 East Francis St.
Ontario, CA 91761
909-930-1751
accufabracing.com

Aeromotive
7805 Barton St.
Lenexa, KS 66214
913-647-7300
Aeromotiveinc.com

Air Flow Research
28611 Industry Dr.
Valencia, CA 91355
661-257-8124
airflowreaserch.com

American Racing Headers
880 Grand Blvd.
Deer Park, NY 11738
855-443-2337
americanracingheaders.com

ARP
1863 Eastman Ave.
Ventura, CA 93003
800 826-3045
Arp-bolts.com

ATI Performance Products
6747 Whitestone Rd.
Gwynn Oak, MD 21207
877-298-4343
Atiracing.com

BluePrint Engines
2915 Cherry Ave.
Kearney, NE 68847
800-483-4263
Blueprintengines.com

Brian Tooley Racing
2049 Filiatreau Dr.
Bardstown, KY 40004
888-959-8335
Briantooleyracing.com

Brodix
301 Maple
Mena, AR 71953
479-394-1075
Brodix.com

CMP
P.O. Box 5385
Sherman Oaks, CA 91413
818-427-3016
Coolmachineperformance.net

Comp Cams
3406 Democrat Rd.
Memphis, TN 38118
901-795-2400
compcams.com

CP Pistons/Carillo Rods
1902 McGaw Ave.
Irvine, CA 92614
949-567-9000
cp-carillo.com

Crane Cams
1830 Holsonback Dr.
Daytona Beach, FL 32117
866-388-5120
cranecams.com

CSU
17367 Reed St.
Fontana, CA 92336
909-851-6955
Csucarbs.com

CXRacing
1627 Chico Ave. S.S.
El Monte, CA 91733
626-575-3288
CXRacing.com

Darton Sleeves
2380 Camino Vida Roble, Bldg. J & K
Carlsbad, CA 92011
800-713-2786
darton-international.com

DNA Motoring
801 Sentous Ave.
City of Industry, CA 91748
626-965-8898
dnamotoring.com

Eagle
8530 Aaron Ln.
Southaven, MS 38671
662-796-7373
eaglerod.com

Edelbrock
2700 California St.
Torrance, CA 90503
310-781-2222
edelbrock.com

FAST
3400 Democrat Rd.
Memphis, TN 38118
877-334-8355
fuelairspark.com

Fel Pro
felpro-only.com

Gandrud Chevrolet
919 Auto Plaza Dr.
Green Bay, WI 54302
800-242-2844
parts@gandrud.com

Holley/Hooker/NOS/Accel
1801 Russellville Rd.
Bowling Green, KY 42101
270-782-2900
holley.com

JE Pistons
10800 Valley View St.
Cypress, CA 90630
714-898-9763
jepistons.com

K1 Technologies
7201 Industrial Park Blvd.
Mentor, OH 44060
616-583-9700
K1technologies.com

Kenne Bell
10743 Bell Ct.
Rancho Cucamonga, CA 91730
909-941-0985
kennebell.net

Lil John's Motorsport Solutions
111 Graystone Ct.
Bardstown, KY 40004
888-583-4408
LilJohnsMotorsports.com

Lingenfelter Performance Engineering
1557 Winchester Rd.
Decatur, IN 46733
260-724-2552
lingenfelter.com

Lucas Oil
302 North Sheridan St.
Corona, CA 92880
Lucasoil.com

Lunati
8649 Hacks Cross Rd.
Olive Branch, MS 38654
662-892-1500
Lunatipower.com

Magnuson Products LLC
1990 Knoll Drive, Bldg. A
Ventura, CA, 93003
805-642-8833
magnacharger.com

Mast Motorpsorts
330 NW Stallings Dr.
Nacogdoches, TX
936-560-2218
mastmotorsports.com

Meziere
220 S. Hale Ave.
Escondido, CA 92029
800-208-1755
meziere.com

Milodon
2250 Agate Ct.
Simi Valley, CA 93065
805-577-5950
milodon.com

Moroso
203-453-6571
Moroso.com

MSD
80 Carter Dr.
Guilford, CT 06437
915-857-5200
Msdignition.com

Precision Turbo
616A S. Main St.
Hebron, IN 43431
219 996-7832
Precisionturbo.net

Procharger
14801 W. 114th Terr.
Lenexa, KS 66215
913-338-2886
Procharger.com

RHS Cylinder Heads
3416 Democrat Rd.
Memphis, TN 38118
877-776-4323
racingheadservice.com

SCAT
1400 Kingsdale Ave.
Redondo Beach, CA 90278
310-370-5501
scatcrankshafts.com

SDPC
5901 Spur 327
Lubbock, TX 79424
800-456-0211
SDParts.com

Snow Performance
1017 Hwy. 24 E., Unit A
Woodland Park, CO 80863
719-633-3811
Snowperformance.net

Speedmaster
1101 W. Rialto Ave.
Rialto, CA 92376
909-605-1123
Speedmaster79.com

Trick Flow Specialties
285 West Ave.
Tallmadge, OH 44278
330-630-1555
trickflow.com

Total Engine Airflow
285 West Ave.
Tallmadge, OH 44278
330-634-2155
totalengineairflow.com

Turbo Smart
8580 Milliken Ave.
Rancho Cucamonga, CA 91730
909-476-2570
turbosmartusa.com

Turnkey Engine Supply
3915 Oceanic Dr., #601
Oceanside, CA 92056
760-966-2663
Turnkeyenginesupply.com

Vortech Superchargers
1650 Pacific Ave.
Oxnard, CA 93033
805-247-0226
vortechsuperchargers.com

Westech Performance
11098 Venture Dr., #C
Mira Loma, CA 91752
951-685-4767
westechperformance.com

Whipple Superchargers
3292 N. Weber Ave.
Fresno, CA 93722
559-442-1261
whipplesuperchargers.com

Wiseco Pistons
7201 Industrial Park Blvd.
Mentor, OH 44060-5396
800-321-1364
wiseco.com